# planet earth

## the making of an epic series

# planet earth

## the making of an epic series

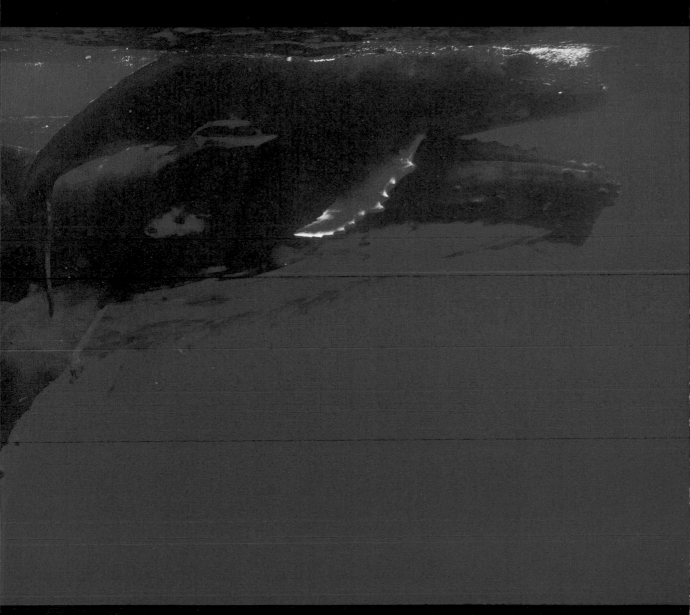

BOOKS

DAVID NICHOLSON-LORD

# contents

# the grand plan
## making a series out of this world

It's probably fair to say there has never been anything quite like *Planet Earth*. This is partly a matter of technology, partly a matter of scale, breadth of vision and budgets, partly because it contains many sequences that have not been filmed before. As a result, it has produced some stunning new views of the Earth, its major ecosystems and the species that live in them. It may even alter the way some of us see the world.

For a series that began as a sequel, that's not bad going. *Planet Earth* was conceived after the extraordinary success of *Blue Planet*, the Attenborough-narrated series on the oceans broadcast at the end of 2001. The series created, in the words of its producer Alastair Fothergill, 'ripples'.

When *Blue Planet* finished, Lorraine Heggessey, who was then controller of BBC1, wanted more – so in January 2002, *Planet Earth* was born. 'My feeling was to do the whole planet in the style of *Blue Planet*,' says Fothergill. But what had made *Blue Planet* so special?

Research showed that viewers liked its novelty – species, locations, sequences they hadn't seen before. They loved its 'epic' scale. And they were impressed by its colour and drama. 'It didn't feel like a lecture,' says Fothergill. 'It was emotional, cinematic.'

*Planet Earth* is a series about superlatives, and much of the subject matter, encountered first-hand, defies the everyday.

Sticking to that brief has meant that *Planet Earth*, too, is epic, dramatic, groundbreaking, 'emotional'. Every programme of the 11 contains, says Fothergill, two or three sequences that have never been filmed before – the aerial footage of wolf or African wild dog hunts, for example, or the astonishing pursuit by a snow leopard of a wild goat on a near-vertical mountain slope in the remote Karakoram range of Pakistan.

Its palette is planet-wide – each separate programme is a package designed to represent not just one mountain range (or desert, or jungle) but all the world's mountains, or deserts, or jungles. It's thus a kind of lexicon of the world's most extreme places – the hottest, coldest, wettest, driest, highest, lowest.

Its makers secured breathtaking high-definition imagery of Mounts Everest and K2 – filmed by the man who shot the aerial scenes for the movie *Black Hawk Down* – and equally breathtaking imagery of the world's deepest and most dangerous caves. One crew filmed over a desert volcano, another in a place of overwhelming wetness – the near-inaccessible Venezuelan table mountain that may have served as the model for *The Lost World*. Cameraman Wade Fairley returned home at the end of 2005 after an entire year in Antarctica recording the dedicated parenting habits of the emperor penguin. Virtually all the crews lived

**For the caves programme, it was, naturally, the world's deepest and most dangerous caves that the team set their sights on – here, Lechuguilla Cave in the Carlsbad Caverns National Park, New Mexico.**

*Overleaf*: In Lapland, a crew took to the air with a cinébulle – a platform beneath a hot-air balloon – so they could drift over the forest without disturbing the snow-encrusted trees.

where they filmed, for weeks at a time – in conditions often as awe-inspiring as their subject matter, and with the experiences to match. These experiences, and their filming successes and failures, are a major theme of this book.

Talk to the people who made the programmes, and the word that crops up most, after 'amazing', 'extraordinary' and 'bizarre', is 'unbelievable'. *Planet Earth* is a series about superlatives, and much of the subject matter, encountered

Virtually all the crews lived where they filmed, for weeks at a time – in conditions often as awe-inspiring as their subject matter, and with the experiences to match.

*Opposite*: Another team spent two months searching for the rare Bactrian camel in the frozen Gobi Desert.

*Below*: Cameraman Wade Fairley spent an entire year on a remote Antarctic island filming emperor penguins.

first-hand, defies the everyday. Thanks to the technology used in its filming, the viewer is able to share more of this than ever before.

The main innovation has been filming in high-definition, which nearly doubles the amount of resolution in the picture and is likened by Fothergill to 'having a stocking over your head that you pull off'. High-definition adds a new dimension of clarity, whether it's polar bears swimming under water in the Arctic or birds of paradise displaying in the jungles of Papua New Guinea; in conjunction with the heligimbal –

a gyro-stabilized camera support mounted on a helicopter (see right) – it has generated aerial footage previously undreamt of in natural history film-making.

There are other important differences from *Blue Planet*. One is the much larger role played by scenery and landscape – in the oceans, for all their beauty, visibility is limited. Another was the pressure on the film-makers – who had less time to make a longer, and in many ways more challenging, series. A third, less easy to describe, has something to do with the way we, the viewers, connect to the images we see.

*Planet Earth* draws attention to the growth of human populations, but it is not a conservation series – not overtly, at least. People don't want to be talked at, it seems. And the BBC has planned other programmes, broadcast alongside *Planet Earth*, that focus on the conservation issues raised by the series. But it will set off resonances nonetheless.

Many of the aerial shots, for example, by withdrawing from a close focus on an individual animal to a panoramic view of its surroundings, give an extraordinary sense of what it feels like to be that animal. For Fothergill, these constitute a 'powerful tool – an emotional way of involving people in a habitat'. And though he talks of the 'relatively small canvas' on which *Planet Earth* operates, the global audience – well over 100 countries – will still be considerable. His chief aspiration for the series is that those who watch it will be 'awestruck by the sheer beauty of our planet'. Like its predecessor, *Planet Earth* will create ripples – and who knows where that may end?

## heligimbals

Gimbals are gadgets used widely in space exploration, aeronautics and defence systems to get maximum accuracy for scanning devices. Developed mainly in the US, they're essentially balancing devices that employ gyroscopes to eliminate wobble. The technology is fairly new, extremely expensive and militarily sensitive.

A heligimbal is a gimbal that goes on a helicopter. It looks like R2-D2 from *Star Wars*, says producer Jonny Keeling, only upside down – a rubbish-bin-sized gizmo bolted onto the outside of the helicopter, usually under the nose. The camera it contains can turn 360° and give a 'rock solid' image from maybe half a kilometre up. Inside the helicopter, there's a joystick to control the camera and a recording deck.

The heligimbal is being used increasingly by news organizations, film-makers and the police; before *Planet Earth*, for reasons of cost as well as novelty, it wasn't used in wildlife films.

# high points and cold spots

The views were the best in the world, but getting to them involved the worst of trips. 'Utterly terrifying' but 'unbelievably exhilarating' are how they describe their highs and lows, from ascending the tallest trees to descending below the ice.

*Opposite*: The world's deepest valley – going for the ultimate shots in the Himalayas. *Left to right*: In magical light – surviving a year in Antarctica. First steps on the ice – close encounters with polar bear cubs. Top shots – getting aerials to take your breath away.

# land of extremes
## chasing wild camels in outermost mongolia

GOBI DESERT Great Gobi
● National Park
MONGOLIA

Travelling by Jeep across the vast, trackless plains of the Gobi Desert, producer Huw Cordey and his colleagues rapidly discovered that two Mongolian phrases were essential to their well-being. One roughly translated as 'I want to go and find my horse,' and the second as 'I want to go and find my camel.' Both related to – how shall we put this? – comfort breaks. Of the two, camel-visiting was the more ... significant.

Visiting one's camel – or one's horse for that matter – is not something to be undertaken lightly in the Gobi. In January 2003, when the team spent two months there in search of the wild Bactrian camel, temperatures dropped to a bitter -35°C (-31°F) at night. Even by day, they never climbed above zero and, with an icy morning wind, could drop back down to -20°C (-4°F). So each evening, when the crew reluctantly abandoned the campfire and crept into their tents for another 10 or 12 hours' shivering, a vital item of equipment was on hand – the urine bottle.

Film-making is sometimes reckoned to be a glamorous occupation. And certainly the Gobi exceeds and occasionally defeats superlatives, at least in terms of its scale and remoteness. Cordey, who has travelled to 50 countries in his career, rates it as 'one of the most exciting and interesting – and also one of the most challenging – filming trips I've ever been involved with'. The key word,

'The experience of wilderness was incredible.
It was so completely otherworldly that it was
almost surreal.'

*Above*: In this vast, unspoilt
place, winter temperatures
never rise above freezing.

*Below*: Field assistant
Tom Clarke carried heavy kit
for hundreds of kilometres
across the Gobi.

perhaps, is 'challenging' – since it's often the most mundane things that put film crews to the test.

Ostensibly, the target of the trip – the wild Bactrian camel – was challenge enough. It is one of the rarest and least-known mammals on the planet, an animal that not only has excellent sight, hearing and smell but also is prone to take off at speed when humans approach to within five or six kilometres (three or four miles) and has been known to run a distance of 70km (45 miles) at a stretch.

The camels are threatened by poaching, drought, wolf predation, interbreeding with domestic camels and, more recently, mange. There are only 800 of them left in the wild, and conservationists worry these could easily disappear without the world knowing.

Then there's the Gobi itself, where the camels survive. At 40,000km² (15,500 square miles), it's an area the size of Holland, with no roads, few people, no settlements. The Great Gobi National Park is the third largest terrestrial park in the world and the flagship reserve of Mongolia, but nevertheless it has no infrastructure, a budget of just $17,000 a year, only one ranger and few tourists. 'As far as I know, nobody had previously spent as long as we did in the Gobi in winter,' says Cordey.

To say the Gobi is a land of extremes may be a cliché, but it's no more than the truth. Cordey describes it as 'a vast, remote, unspoilt wilderness. The logistics of filming there were almost impossible. In effect, it's outer outer Mongolia. For six months of the year it's freezing, and for the other six months it's boiling.'

For the two-month expedition, the team travelled in four vehicles and took with them all their food, fuel and water. It was an all-male trip – Cordey, cameraman and scientist Henry Mix, assistant cameraman Tom Clarke and conservationist Richard Reading, of Denver Zoo. Washing was limited to stores of wet wipes. Shaving was out of the question. So, too, was alcohol. Even so, at least one fight broke out among the drivers after they had got hold of vodka.

The local food was notable for the ubiquity of mutton, the absence of vegetables and the popularity of fat. 'The drivers were always asking if they could eat our fat as we Westerners carefully picked our way around it – at least that's what we did at the start, before we realized that we would probably die of cold if we didn't eat the stuff,' says Cordey. And then there was the Great Gobi night.

'The days were glorious, but for me, the nights were utterly miserable,' Cordey recalls. 'I couldn't seem to keep warm at all. I don't think I've ever had more clothing on – maybe five or six layers. As the night closed in, I started thinking about having to go back to my tent. I used to dread it. We would literally sit on the fire. I burnt so much clothing – gloves, the seat of my pants. But you could only warm what was in front of the fire – you would have a hot thigh, say, but a freezing backside.

'The days were glorious, but for me, the nights were utterly miserable,' Cordey recalls. 'I don't think I've ever had more clothing on – maybe five or six layers.'

'In the morning you'd wake up, and the whole of the inside of the tent would be glinting with little ice crystals. Absolutely everything liquid that you didn't sleep with would freeze. The wet wipes froze in seconds, and so you had to sleep with those. I put my contact lenses in my pocket and slept with them there. Clarkie [assistant cameraman] had to sleep with all the batteries. All of us were going to bed with a ragbag collection of items. It was the same with the urine bottles. If you didn't keep them in your sleeping bag with you, they froze. On one occasion I didn't, and I then had to spend about half an hour unfreezing my urine over the campfire.'

As for the camels, 'it involved huge amounts of walking and driving just to get near them. We got lucky only when absolutely everything was working in our favour – terrain, wind direction. You would see them in the distance running over the horizon – they weren't stopping. But in the end, we did get some shots of truly wild camels.

'The whole thing, for me, was one of the highlights of my filming career. There were times when I lost all feeling in my fingers, but the experience of wilderness was incredible. It was so completely otherworldly that it was almost surreal. It was tremendously liberating.'

*Above*: Domesticated camels are vital to the local people, who use them for transport, meat, wool and milk.

*Top*: The campfire provided precious warmth in daytime temperatures down to -20°C – and was also useful for defrosting urine bottles.

*Left*: Riding a camel 'was rather like sitting in a very comfortable old armchair', says producer Huw Cordey.

# beyond the edge
## eye to eye with everest

F lying in a military plane at 8,500m (28,000 feet) with the aircraft door open is not high on the list of recommended cinematographic health-and-safety procedures. Nor is crawling around near the door to fix a frosted-up camera – or dropping 3,000m (10,000 feet) in about 10 seconds when one of the air crew has a sudden and life-threatening bout of altitude sickness. It wasn't meant to happen that way, of course. But when you're going for the ultimate shot, the stakes are high.

The footage of Mount Everest and K2 filmed for the mountains programme broke several records for high-altitude filming and flying and provided a clarity and stability of image that is probably without precedent for the two peaks. The shots were taken using 'heligimbal' technology (see p9) – gyroscopic balancing mechanisms that cut out wobble. But in both cases, it meant what producer Vanessa Berlowitz calls 'operating at the limits of what's safe'.

The use of the heligimbal is one reason why the shots of Everest and K2 are so impressive. Another is the height at which they were taken, ranging from 6,700m to 8,500m (22,000 to 28,000 feet) – virtually on the same level as the summits. Yet another is the cooperation of the Nepali and Pakistani armed forces, without which the sequences would have been impossible. Berlowitz and cameraman Michael Kelem flew in a variety of military planes and helicopters – including, in Nepal, a Skytruck aircraft also used for fighting insurgents.

The negotiations with both countries took more than a year, relied heavily on the global reach of the *Planet Earth* series and meant that the producers had

'It's very risky flying ... You get these strange backwashes: you can be flying close to a ridge, and then the air drops off – you get a void, and the helicopter can literally plummet.'

To film clouds building up in the valley with Everest as a backdrop, the team – with 35 porters – trekked for a week to Gokyo in Nepal. They made the final ascent seven times to get the perfect conditions for the shoot.

access to the best pilots in the world – pilots uniquely skilled in some of the most difficult flying conditions imaginable. For the two countries, both with political problems, it would serve as a kind of international shop window, reminding the world of just how awe-inspiring their landscapes are. More pragmatically, both regions probably could not have been accessed without government help.

'You can't just ring up and hire a helicopter,' says Berlowitz. 'These places don't have commercial helicopter agencies. And it's very risky flying. There's nowhere to put down when you're flying round a peak, and at that altitude, you wouldn't be able to take off again anyway. The high-altitude winds are such that you need very experienced pilots to understand all the dynamics. You get these strange backwashes: you can be flying close to a ridge, and then the air drops off – you get a void, and the helicopter can literally plummet.'

'He was sitting opposite me, and I noticed his fingers trembling and twitching ... his eyeballs started to roll backwards.'

The team relied on the highly skilled crew of a Pakistani military helicopter to fly into the Karakoram range. The helicopter was fitted with a bracket for the 'heligimbal' before departing from Skardu army base.

The Karakoram range, which includes K2, is probably more hazardous than the Everest region, because the peaks are clustered closely together, heightening the unpredictability of the winds. The shoot also involved overflying the Baltoro Glacier, part of the largest glacier system in the world outside the poles. Two helicopters went, to provide immediate rescue in case one crashed. 'No-one had done aerial filming up there before, and the army was very worried about how the camera would affect the aerodynamics of the helicopter at altitude. The other crew flew ahead and checked the wind movement and the safety of the flight path. It was a huge extra expense, but we were very grateful to have that second helicopter.'

For both individuals and technology, conditions were highly testing. Explains Berlowitz: 'In the past, people have shot from planes but always from wobbly cameras and through a window. We wanted to have the door open and to use this fantastic piece of equipment, the heligimbal. We were trying to push boundaries and increase the quality of the material to a level that hadn't previously been achieved. It was very ambitious.'

This had two consequences. First, the filming equipment, including the heligimbal, was being used in circumstances for which there was no precedent.

And second, the cabins could not be pressurized, and so the flying and filming crews had to wear oxygen masks and cylinders and full polar clothing. They also had to communicate through voice-coms in their masks – a potential source of misunderstanding.

The first flight in Nepal was in a specially modified high-altitude helicopter, the Squirrel B3, which took the film crew to 7,000m (23,000 feet), thought to be a record for the helicopter. The shots were impressive – but they still weren't summit-level. So the Nepalis, despite being preoccupied with Nepal's long-running Maoist insurgency, agreed to go even higher, and to fly Berlowitz and Kelem to 8,500m (28,000 feet) in a Skytruck.

'I felt rather guilty about taking them out of a war situation to do film footage, but they said it was a diversion for them – almost a holiday. The day we filmed, they had been on an early-morning mission against the Maoists. So at about 4am, they were unloading mortars and shells and bomb casings and loading our equipment in. We took the door off and positioned the heligimbal in the doorway. We were going to fly from the Chinese side of Everest over the edge. The weather was good, and they told us we had a 25-minute window – that was all. So we

Cameraman Michael Kelem had just one chance to get summit-level shots of Everest. At 8,500m, in a Nepali army Skytruck plane with the door open, all on board had to wear oxygen masks and polar clothing.

Approaching Karakoram's Trango Towers, the crew prepared to fly among the densely clustered peaks. This area is more hazardous than the Everest region – and arguably even more breathtaking.

would have one or two chances at the shot. It was very pressurized.'

The first sign of something wrong was the monitor inside the plane relaying the pictures back from the heligimbal. 'The door was open, and so it was absolutely freezing. Michael and I could barely see each other over our oxygen masks. We lined up for the shot of Everest, and I looked at my monitor – it seemed soft. We were talking to each on the voice-coms through the oxygen masks, and I said: "What the hell's happened?" He said we'd got frost on the front of the lens.

'Michael was fantastic. He crawled on his hands and knees, swivelled the camera to face back inwards, got an electric screwdriver, undid the front of the camera system and scraped off the frost. He had to take his gloves off – I could see him shaking in the cold. He was close to the open door, he wasn't on a harness and he had only minutes to do it before we had to descend and lose possibly our only opportunity at the shot. It was an extremely tense moment. Fortunately, we had a Nepali engineer on board who was helping him and holding his oxygen cylinder.'

With the camera cleared, Kelem got his shot, and there was just time for a second run – which the pilot agreed to. 'We went back round and positioned ready for the shot, but as I was looking at the monitor, I happened to catch sight of the engineer out of the corner of my eye. He was sitting opposite me, and I noticed his fingers trembling and twitching. I have worked quite a lot in mountains, and I recognized it as a symptom of hypoxia – altitude sickness.

'As I looked at him his eyeballs started to roll backwards, which is a quite advanced symptom. His oxygen mask was on, and I asked him if he was OK, but he wasn't responding, which at that altitude means you've got minutes – it's potentially fatal. I shouted in the intercom to the pilot, the co-pilot leapt back and we realized the engineer's oxygen supply wasn't working. The co-pilot started sharing his oxygen with the engineer – at which point the co-pilot said his oxygen supply had had it, too. I shouted at the pilot that we had to descend – so within minutes, we had gone from being lined up for a beautiful shot of Everest to a near-fatal situation.'

The descent that followed, says Berlowitz, was 'unbelievable – unbelievably painful and frightening. It was about 3,000m [10,000 feet] in ten seconds. Our eardrums kind of exploded. But Michael was further down the plane, still trying to film. He couldn't understand why we had suddenly dropped down through the sky and why his shot was all over the place. I was connected through the voice-com to the pilot and the engineer, but Michael was just connected to me. He had no idea what was happening.'

Back on the ground, they worked out what had gone wrong. The engineer must have lost a valve on his oxygen cylinder while helping Kelem repair the camera. Responses differed – the other Nepalis teased the engineer and made light of it, the man himself hugged Berlowitz and told her she'd saved his life.

The thought that he was probably right was a sobering one. Filming later around K2, on oxygen again because of the altitude and with both the weather

'We got great material, but then a massive weather front moved in, and the camera kept crashing – the electrics went. It was pretty heartbreaking ... maybe it was a sign.'

and the filming timetable closing in, there was a heightened sense of what could go wrong. 'We'd already had to wait for our permit to come through, and so we'd missed ten days of beautiful weather. We flew round and round some of these extraordinary spires and structures, down the gaps between the towers, and it took my breath away. I thought the Everest region was amazing, but the Karakoram range knocked the spots off it. It was absolutely the most extreme mountain environment I have ever seen. We got great material, but then a massive weather front moved in, and the camera kept crashing – the electrics went. It was pretty heartbreaking. So we landed and said maybe it was a sign – let's get out of here.

'I felt very shaken after filming Everest. It was a bit of a wake-up call for me. We're all used to risk, and there's a lot of pressure on a series like *Planet Earth* to get the ultimate shot, a shot that no-one has achieved. I suddenly thought how I would have felt if we had got the shot but at the cost of someone dying. I do occasionally wake up at night and think: "If only we'd had another couple of days at K2." But we got what we got. Maybe I just have to be grateful we got up there at all.'

## the highest ever pizza delivery

The *Guinness Book of Records* doesn't contain a section on high-level pizza delivery – but maybe it should. Researcher Jeff Wilson and cameraman Gavin Thurston were filming migrating demoiselle cranes in Nepal for the programme on mountains. Says Wilson: 'The hostel in which we were staying noticed that we had forgotten to take any food with us as we climbed up the sides of the deepest valley in the world to film the cranes migrating through the Himalayas. Not long after we'd struggled to an altitude of about 14,500 feet [4,400m], a local guy turned up with two steaming hot pizzas wrapped in tin foil for our lunch. He'd come from the bottom of the valley.' They reckoned it set a world altitude record for pizza delivery.

# let battle begin
## frustration with the head-bangers

GREENLAND
*ARCTIC CIRCLE*

Kangerlussuaq

Musk oxen are among the champion head-bangers of the natural world. From 100m (330 feet) apart, the big males smash into each other with a ferocity that echoes round the hillsides of the tundra. They're built for it, too – reinforced skulls, super-thick horns. So if you went to watch them for a month or so during the Arctic summer, when their hormones are running high, you'd be bound to see some awesome, shattering, high-decibel conflicts.

For the team making the programme on the polar regions, it was almost a no-brainer. To ensure their success, they had sought the advice of musk ox researchers, who suggested a broad valley in the southernmost part of the Arctic Circle in Greenland – where head-banging musk oxen were ubiquitous. Fights took place 'five or six times a day'. The crack of jarring crania was commonplace.

The valley lies about 30km (20 miles) northeast of the airport at Kangerlussuaq. A three-person team – cameraman Martyn Colbeck, producer Jonny Keeling and musk ox researcher Katrine Raundrup – were dropped off by helicopter in the summer of 2003. Keeling recalls: 'We didn't think there would be any problems. We'd just follow the males around and get loads of head-bangs. We might even manage slow motion. It would all be very impressive.'

The team stayed in a small hut used by researchers, from which, every day, they set out across the tundra with their gear. 'We got into a good rhythm,' says Keeling. 'It's a big, broad open valley, 25km [15 miles] long and about 10km [6 miles] wide, with the Greenland ice cap on one side and hills on the other. There were herds of 10 to 20 musk oxen drifting through – you might get four herds in a day. Each morning, we would look for the herds and pick out ones with big males in them or nearby. The male in a herd would be defending his females; the other males might challenge him. We walked 20 to 25km [15 miles] a day, following the herds, and then walked back to camp. We'd cook our meal, get up the next morning and do the same thing.'

As the days turned into weeks, the team realized that head-banging was not high on the musk-oxen's agenda. Nevertheless, as the peak of the rutting season neared, there did seem to be a gradual increase in intensity. And there was a wealth of scene-setting imagery to film.

'We got all the establishing shots, all the build-up, the mating with the females, which is quite rare to get on film, the images in lovely light, of fur billowing in the wind, the beautiful scenics – we got absolutely everything. We thought we were building up to some big fights. The males would paw the ground like cartoon bulls, rub their heads on the earth, circle round each other, posturing and bellowing and roaring. It could be quite unnerving. They didn't mind us as long as we were downwind and kept still. With the wind swirling around, we had to be careful. Sometimes, when the males caught sight of us, they would come marching over,

*Opposite*: With reinforced skulls and thick horns, male oxen are well adapted for crashing their heads together in fierce battles over females. But if fighting can be avoided by posturing and bellowing, then it is.

*Below*: The Arctic summer in west Greenland – rutting season for musk oxen – promised some spectacular action between rival males. It was just a question of choosing a location, then following the oxen around.

thinking we were females to be rounded up or other males to be challenged. They'd be bellowing the whole time – an incredible sound, just like a lion's roar, which seems to resonate in your rib cage, particularly if you're close to them.

'There's nowhere to hide on the tundra, no rocks or trees. It's completely flat, and you're stuck behind a little carbon-fibre tripod, crouching behind each other. There's this animal walking towards you that looks fairly benign, a kind of giant sheep, like a big Dougal from *The Magic Roundabout*. Then it gets close and starts bellowing, and suddenly it looks like a very ferocious buffalo. People have been killed by musk oxen charging. But you just have to stand there and hope he realizes you're human. Some of them got as close as 5 to 10m [30 feet]. And then, at the last minute, they'd realize you weren't a musk ox and run away back to the herd.'

Not a single fight took place in three weeks. So at the end of the third week, the team got the go-ahead to stay on for a fourth. But the last day dawned with the male oxen still bellowing and posturing. 'We had managed to get a shot where a pair of males dropped their heads about a foot away from each other and sort of tapped their heads together, but it wasn't very impressive at all.' And

The crew watched and waited, here, as a caribou wandered past. They filmed all the other shots they needed – including some rare footage of musk oxen mating – but the key fight sequence proved elusive.

'It would have been the perfect shot to complete the sequence. And there was absolutely nothing we could do ...'

two males had begun squaring up to each other – before rushing out of sight over the brow of a hill to consummate their collision. 'We heard the crack, but it was just behind a fold in the landscape, and so we couldn't film it,' says Keeling.

That was all the action they had seen in four weeks. The last day began very quietly, just like every other day. The crew were packing up to leave – with the camera gear stashed away in boxes – when two males lumbered into view, pawed, snorted, bellowed – and charged. 'It was the most almighty fight,' recalls Keeling. 'They charged at each other probably three times. The light was beautiful. All it required was 30 seconds of footage. It would have been the perfect shot to complete the sequence. And there was absolutely nothing we could do – it was all over too quickly. We just could not believe that, on the last hour of the last day, they had suddenly decided to fight in sight of our camera – if only it hadn't been packed away.'

Why weren't the musk oxen fighting? Probably because they didn't have to. 'Most animals don't want to fight; they just want to shout and posture to decide who's the boss,' Keeling points out. In this case, the suspicion was that meat- or trophy-hunters had visited the valley the previous winter and killed many of the larger males, skewing the sex-ratio and reducing the males' need to compete.

'I felt physically sick at the time,' says Keeling. 'I went on feeling awful for about a week ... We did our best – we worked as hard as we could every day. It was just bad luck. Sometimes you get the shot on the first day. Sometimes you don't get it at all.'

**Up to four herds a day of 10 to 20 musk oxen came through the valley. Images of their Dougal-like fur billowing in the wind added to the mix of shots that could be used in the build-up to a fight.**

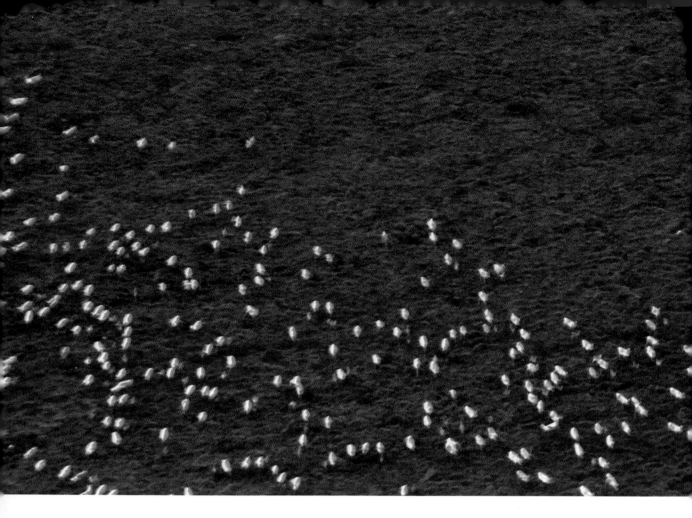

# magic in the wilderness
## going hi-tech for the wolf hunt

ARCTIC CIRCLE

● Lupin

**CANADA**

Ask most people if they've ever seen a successful wolf hunt on television, and they'll probably do a double take. Wildlife films are full of them, aren't they? In fact, there's usually some clever illusion-making involved. We see wolves running, we see them eating their prey. Maybe we see them dragging it down. But are they the same wolves filmed at the same time? And what about the hunt itself – are we seeing the whole thing from start to finish?

Producer Jonny Keeling can think of only one film that depicts a successful wolf hunt from above – the only place from which the entire hunt can realistically be seen. It's an old sequence, shot from the open door of a Cessna light aircraft, and as 'wobbly as hell', he says. Thanks to the arrival of the heligimbal in wildlife film-making (see p9), and its use over the tundra of northern Canada in the *Planet Earth* series, we now know what a wolf hunt really looks like – a whole hunt, the same wolves throughout – and there's not a wobble in sight.

The team that went to Canada didn't set out to film a wolf hunt. Biologist Anne Gunn, an expert on the wolf-caribou relationship, told them they wouldn't manage

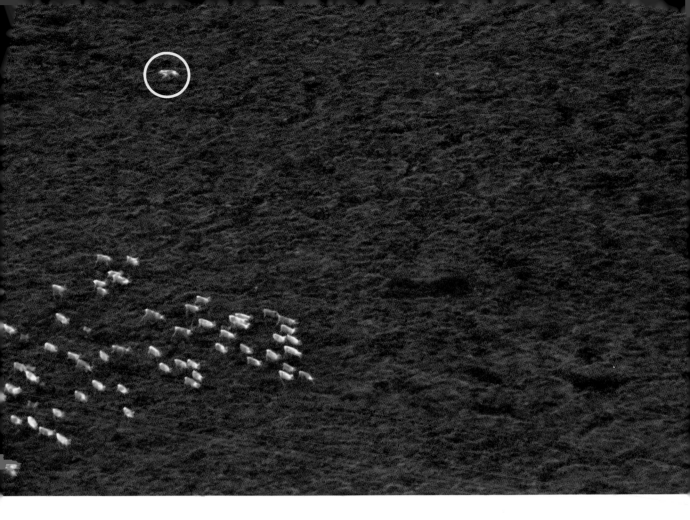

The team that went to Canada didn't set out to film a wolf hunt. Biologist Anne Gunn, an expert on the wolf-caribou relationship, told them they wouldn't manage it – the odds were too low.

it – the odds were too low. Cameraman Jeff Turner, who had just finished an hour-long BBC2 programme on wolves, was of the same opinion. He'd been filming at Bathurst Inlet in Canada, where the crew followed the animals by float plane, landing on the nearest lake when the wolves were spotted circling a caribou herd. Turner would jump out and walk four or five kilometres (three miles) across the tundra carrying all his gear – often to find the wolves had vanished, or the chase had taken them much further on.

It was not even certain the *Planet Earth* team would get the 50,000-strong concentrations of caribou needed for good aerial shots. Much depended on the weather. Warm, still weather brought out the flies and mosquitoes, which made the caribou band together – and produced 'beautiful patterns' for filming, says Keeling. Cold weather meant far fewer insects, more strung-out caribou herds and much less impressive images. But at the very least, Keeling thought, the heligimbal would produce 'a sense of what it's like for the animals to live in this massive open space'.

In the event, the weather was cold. A five-strong team helicoptered out in July

When wolves (circled in white) are hunting caribou over the tundra of northern Canada, they can cover such a lot of ground – sometimes up to 10km – that the only way to film an entire hunt from start to finish is from the air.

2004 to Lupin, an isolated gold mine in the badlands of the north Canadian tundra, where they lived alongside 100 miners in an encampment of prefabricated huts. Each day, guided by satellite collars on the animals and GPS coordinates supplied by Anne Gunn, they flew more than 100km (60 miles) north into the Arctic Circle to find the caribou herds. As Gunn had predicted, the caribou were thin on the ground and the wolves elusive. Then the team's luck turned, dramatically.

Keeling recalls: 'It was about the fifth day. We were out in the helicopter; it had just started to rain, and we were about to refuel before going back for the night. It's normally very difficult to see animals on the tundra from a helicopter, but the wolves were white and so easier to spot, and we saw a group of them trotting. Just ahead, there was a big bunch of caribou. The wolves looked purposeful, and Jeff said he thought something was going to develop.

'We put him down on the ground with the caribou and then took off back to the wolves. By the time we got there, we'd completely lost sight of him. The herd of caribou, even through binoculars, were a grey smudge on the landscape. They must have been three or four kilometres [two or three miles] away, but the wolves started to run. They were making a beeline for the caribou, and we began filming them with the heligimbal.

'We had never used the system to film animals before. The wolves weren't bothered at all. They latched onto a young caribou and chased it towards where we had left Jeff. They caught and killed it right in front of him. It was extraordinary ... on our very first attempt, we got shots from the ground of the very same hunt that we were filming from the air.' Even more fortuitous was the timing: 'about 20 seconds' after the kill, the heligimbal broke down. 'We had to land and check it. It was burnt out, but we managed to repair it on the ground.'

The next few days were even better. 'If we landed quite close to the wolves, we

Most animals on the tundra are hard to see from above, as they blend in so well. But the wolves were white and so easier to spot. Their stamina was remarkable – the team saw one flat-out chase lasting ten minutes.

At least, Keeling thought, the heligimbal would produce 'a sense of what it's like for the animals to live in this massive open space'.

could watch them with binoculars until they got into hunting mode. We would take off and fly over them until they caught and killed something. Then we'd land and wait for them to do it again. It was great that we could spend a long time with them – there were 24 hours of daylight – and we could save money by waiting on the ground instead of flying.'

In the five or so days that remained, Keeling and his colleagues managed to film seven separate hunts, from start to finish, some covering up to ten kilometres (six miles), one involving a flat-out chase that lasted ten minutes. The potential of the new technology began to dawn on them.

'It wasn't just the hunts themselves – you also get this fantastic bird's-eye perspective, with the caribou fanning out left and right, making wonderful patterns, and the wolves running in among them. You realize what the wolves are having to cope with. At the end of one of the hunts the camera pulls out from the kill and reveals this absolutely immense wilderness. When you see that image, you really begin to appreciate the animal and the environment it lives in.'

The sequence was so compelling it was promoted to the opening programme of the series. Keeling adds: 'After that first hunt, we got back to the mine and sat down for a meal – and we just looked at each other. I think we suddenly realized what an incredible filming system it was and what opportunities it opened up. It was one of the few filming trips I've been on where you come back with something far greater than you ever dreamt of.'

*Above*: For filming aerials, the heligimbal – an anti-wobble device containing a camera – was attached to the front of the helicopter.

*Below*: The team wanted aerial images to convey the immense wilderness that the wolves live in – and the heligimbal proved an ideal way of capturing them.

# spirits of the north
## tracking white wolves in the snow

ARCTIC CIRCLE
Banks Island
CANADA

The visit by *Planet Earth* to the Arctic in April 2005 set a number of precedents. It was the first time the BBC Natural History Unit had been to Banks Island – an Ireland-sized lump in the far north of Canada. It was the first time they had filmed genuinely wild Arctic wolves. And it was the first time researcher Justin Anderson had ridden a Ski-Doo, seen sastrugi or played Boggle (a kind of slimmed-down Scrabble) in a tent in temperatures down to -35°C (-31°F) while going 'stir crazy' waiting for a five-day white-out to finish.

A Ski-Doo is a snowmobile, and sastrugi are wind-blown ice sculptures the Ski-Doo-rider must negotiate. 'It's a bit like driving over a Christmas cake,' says Anderson. Riding back to camp as the storm struck, peering through frozen goggles into driving snow, he tried a short cut, crashed into a large hole in a creek bed and flew over the handlebars. He wasn't hurt – just embarrassed.

The snow that cushioned his fall was one reason the team was driving up to 50km (30 miles) a day without finding wolves. They had the help of two of the last traditional hunters among the western Inuit, John Lucas and son Trevor, but even the Lucases had to rely on GPS units as the white-out struck. The real problem, however, was that the landscape was snowy white – and so were the wolves.

Previous films on the Arctic wolf had used a pack habituated to humans through feeding by scientists. *Planet Earth*'s aim was to film unhabituated wolves hunting musk oxen, for a section on the polar spring, 'though we knew realistically that this was like looking for a needle in a haystack,' says Anderson.

Banks is home to about 250 Arctic wolves, but there's 70,000km$^2$ (27,000 square miles) of island over which to find them. Fortunately, there are also some 35,000 musk oxen, which are 'the only dark things for miles around'. The oxen serve as a focus. Other species gather round to exploit the holes they break in the snow; the wolves target their newborn calves. Cameraman Simon King spent hours stretched out on the snow letting the oxen get used to him. But when wolves finally appeared, the timing could hardly have been worse.

'We were waiting for a plane to arrive with food and gasoline ... when John [Lucas] suddenly came hurtling through camp on his Ski-Doo, gesticulating wildly, before riding off towards where Simon was filming the musk oxen. Through binoculars, we saw there were four wolves heading in the same direction. We tried to raise Simon on the radio, but we had to whisper in case we scared the wolves. It was unbelievable. It was the first time we'd seen wolves, and any moment the plane was going to land and the animals were probably going to run a mile.'

But the wolves found King, who filmed them 'coming towards him and having a good look and a sniff around', catching lemmings and finally feeding on a calf they had killed. 'That was the last we saw of them,' says Anderson. 'We kept coming across dead calves and wolf-prints the size of hands, but we were always one step behind. They were wild wolves doing their own thing. If they came and looked at us, it was because they wanted to, and when they'd had enough, they disappeared again. I think we were incredibly lucky to have see them at all.'

*Opposite top*: One way to find a white wolf after a white-out is to hang around its prey. Simon King spent hours on the snow letting the musk oxen get used to him. His reward was oxen within 20m (65 feet) – then a wolf 15m (50 feet) away.

*Opposite bottom*: The team saw no-one during the five-week trip other than the plane crew who brought supplies after the storm. But the noisy arrival coincided with Simon King's only chance to film the wolves, just 2km (1.2 miles) from the camp.

# up, up and beyond
## climbing the world's tallest trees

CALIFORNIA

Jedediah
Smith Redwoods
State Park

USA

**Earendil is one of California's giant redwoods – the tallest organisms in the world, growing to more than 100m (330 feet) high. The reward for scaling such a height was the sight of a garden of smaller trees and shrubs and, living there, a population of animals.**

At more than 100m (330 feet) in height – taller than the clock tower that houses Big Ben – the giant redwoods of northern California are the loftiest trees on the planet. 'When you stand below them and look up, you can't believe they're so big,' says assistant producer Penny Allen. But, as every amateur photographer knows, it's one thing to find a view impressive: it's quite another to convey this on film.

To give a sense of the sheer grandeur of the redwood forest of California, a sense of scale was needed – something to measure the trees against. The result, for Allen at least, was one of the most exhilarating moments of her life.

Producer Mark Linfield and his team decided they needed a tracking shot – a continuous sequence of film that would convey both the height of a tree and the scale of the forest canopy. In effect, the camera would be climbing the tree in one continuous movement – the vertical tracking shot to end all vertical tracking shots. The sense of scale would be provided by climbers stationed at various points on the tree trunk, overtaken by the camera during its vertiginous ascent.

The shot took five days to set up, with an eight-person team of camerapeople, specialist climbers and scientists festooning one of the world's tallest trees in a complex of ropes, rigging and cables. A vital ingredient was the cable dolly developed by cameraman Ted Giffords, which enabled the camera to be transported vertically or horizontally. Up in the canopy, meanwhile, a 'Tyrolean traverse' was set up, 70m (230 feet) across and 90m (300 feet) above the ground, connecting the largest living redwood to one of its giant neighbours.

The work involved was complex and laborious – walking on foot into the forest each morning with filming gear, spending 14 or 15 hours a day there to maximize use of daylight, firing crossbow bolts with lines attached up into the tree, looping the ropes and cables round branches. It didn't help that the lowest usable branches were about 60m (200 feet) above the ground.

For Allen, the highlight was the one occasion she crossed (in harness) the Tyrolean traverse. Like every other climbing member of the crew, she had gone through a week's industrial rope-access training; she is also a rock climber. But neither had quite prepared her for what was 'possibly the most exciting experience of my entire life ... my hands were shaking so much that most of the photos I took at the beginning of the traverse were too blurred to use. I'm not really bothered by heights, but there was an amazing sense of exposure – partly because you're trusting your life to a branch. And 90m is a long way to fall.' It was, she says, 'utterly terrifying but unbelievably exhilarating'.

A vertical tracking shot on this scale had not been attempted before, but it worked so well that it forms the opening sequence of the programme. There's also footage of a climber on the traverse – shot from a heligimbal (see p9). The zoom lens focuses full on the climber, then pulls further and further out. 'In the end, you've got this tiny grain-of-rice-sized person in among the giant redwood forest,' says Allen. 'It really does give you a fantastic sense of perspective.'

# beneath a frozen crust
## exploring the world's deepest lake

SIBERIA

● Lake Baikal

MONGOLIA

Filming *Planet Earth* was partly about filming superlatives – the natural spectacles that set the records – and there's no doubt Lake Baikal falls into that category. It's the world's deepest lake, and also its most capacious – it holds a fifth of the world's surface fresh water. And because it's so large and so cut off, it has developed a bigger range of endemic species (species that are found nowhere else) than any other freshwater lake.

Baikal, stretching 640km (400 miles) across Siberia just north of the Mongolian border, has the world's only freshwater seal – the nerpa. It has forests made of sponge. It has a vast number of invertebrates, including the voracious 9cm-long gammarid shrimp, a scavenger which gobbles up just about any dead matter it comes across. It also has ice.

Baikal is ice-free in summer but starts freezing over in October. By midwinter, it is sealed off beneath a crust at least a metre thick. Around April, it starts to melt again, and the crust of ice that covers the lake begins breaking up into huge plates. It's probably the riskiest moment in the year to traverse the lake. Naturally enough, it's the moment *Planet Earth* chose to film there.

Why would anyone spend a month driving across the melting ice of the world's deepest lake in a couple of battered old Russian camper-vans? There are two answers – one simple, the other less so.

The simple answer is that you need two camper vans in case one gets stuck or, worse, falls down between the ice plates into the waters below – which are 1,800m (more than a mile) deep. To be fair, if that happened, 'you'd be a goner', says producer Mark Brownlow. But at least there would be someone to raise the alert.

Filming under water in Baikal as the ice is melting is not for the faint-hearted. You pick your spot for the dive and then drive out across the ice. Sometimes there's been a snowfall the night before, and so you have to prod around with ice picks for firm ground. If you come to a gap between ice plates, you lay planks across it to form an impromptu ramp and then wait as the driver revs up and shoots across.

Perhaps most unnerving is venturing out onto the lake in the cool of the morning but having to return in the afternoon after a sunny day. 'You get these cannon-shot sounds from across the lake as the plates begin to melt and crack,' says Brownlow. 'You're on tenterhooks the whole time.'

Once you've driven to your dive site, you have to get out the ice picks again – and spend up to an hour cutting yourself a hole through the ice sheet. Then you put on your dry suits and air cylinders – and make sure, once you're under the ice, that you know how to find that hole again.

The more complex reason behind *Planet Earth*'s decision to film in Baikal during the ice melt is because that's when the lake's submarine landscape is at its most stunning. In winter, a blanket of snow cuts out the light and makes the water below pitch black; in summer, algal blooms turn it green and limit visibility. Timing was critical: diving as the snow and ice melted but just before the ice broke up meant not only good visibility but also the chance to photograph the lake's dramatic

*Left*: Once out on the lake, it could take up to an hour to cut a hole in the ice before the dive could begin. But it was so cold that the divers would stay under water for 45 minutes at most.

*Below:* The small ice hole was the only way out. Lose it, and you would suffocate under water. One diver remained on a safety line at all times, and it was vital to stay in contact with him.

For underwater filming, the camera was sealed inside a protective housing, but the extreme cold caused the seals to leak. It also meant the camera couldn't pick up colour properly, the images losing all but the blue tones.

underwater ice sculptures – and maybe also get closer to the seals. Nobody, says Brownlow, had filmed them under the ice before.

Diving under frozen Baikal is not much less intimidating than driving over it. The water is only just above freezing, and so it's a question of 'how long you can last under water before the pain gets too much ... I have never been in extreme cold like that before. The main sensation is burning – intense heat on your lips and on your face. You've got about 45 minutes, max, before you've had enough.'

Safety precautions had to be elaborate. Boiling water was poured over the air

'If you lose your position you're never going to find your air hole again,' says Brownlow, 'and then you're stuffed, because you can't break through the ice from underneath.'

cylinders and regulators at the start of the dive to stop them freezing under water. The team used 'rebreathers', which recirculate air in a closed system. The carbon dioxide absorbent keeps the diver warmer; the absence of bubbles makes him soundless and therefore, in theory, able to get closer to the seals. But rebreathers can kill if rigged up wrongly, and so a rebreathing instructor came along, too.

One diver remained on a safety line throughout, and it was vital to stay in contact with him. 'If you lose your position you're never going to find your air hole again,' says Brownlow, 'and then you're stuffed, because you can't break through the ice from underneath.'

The freezing temperatures nevertheless 'increased the cock-up factor exponentially'. The protective camera-housing leaked, cables froze and snapped, short-circuiting electronics. There was 'a lot of circuitry going on in the evenings, in between vodka shots'.

The seals, which are hunted locally and thus nervous of humans, proved more difficult to film than envisaged. But the extraordinary submarine world of Lake Baikal lived up to expectations. Adds Brownlow: 'I've never ice-dived before, and a lot of people said I was mad, I was going to hate it. They couldn't think of anything more claustrophobic – the idea of going through a small hole and knowing that, unless you find that hole again, you're going to suffocate in this sealed underwater world.

'It didn't bother me at all. It was an incredible experience. Baikal has these amazing translucent ice plates under the water, which you don't get in the Arctic or Antarctic at all. Then there were the seals and the bizarre sponge forests and the giant, primeval-looking gammarid shrimps. You kind of pinch yourself, because you think you're in the sea. It was unbelievably beautiful – an underwater landscape that looked very different from anywhere else in the world.'

Some cracks in the ice could be bridged with a couple of planks, the driver revving up the engine and then shooting across. But a route that was passable in the cool early morning might no longer be safe after a sunny day.

# living like emperors
## filming in the worst winter on earth

Macey
Island
MAWSON ●
ANTARCTICA

The emperor penguin has an unbelievably tough existence. It's not only that it breeds in winter: it breeds in just about the worst winter on Earth. Antarctic temperatures plummet to -60°C (-75°F), blizzards rage at up to 200kph (125mph), freezing katabatic winds blow off the polar plateaux, the sun is, for months on end, a memory. It's an environment that's dark, absurdly cold, brutal and often hazardous – but if you want to film 'emps', there's no alternative. You have to share it.

The stunning sequences of the emperor penguin breeding cycle filmed for *Planet Earth* were possible only because cameraman Wade Fairley and biologist Frederique (Fred) Olivier made Antarctica their home for an entire year. They lived for much of this time in a field hut the size of a garden shed and with the

Insulation properties of 'a lump of cold steel'. It was probably just as well that they both loved birds, remoteness, filming and the strange, twilit magic of a sunless polar winter.

The hut in which Fairley and Olivier spent their Antarctic winter is on Macey Island, a small rocky outcrop about 60km (40 miles) from Mawson station, one of the bases of the Australian Antarctic Division (AAD). For the filming, AAD offered *Planet Earth* extensive logistical support – accommodation, transport and equipment. Mawson is well established – the oldest station south of the Antarctic Circle – but the team were to spend most of their time on Macey.

Macey hut is a refurbished 1950s freight container, which shakes violently in a blizzard – like being in an earthquake zone, says Fairley. For lighting, the team often had to rely on candles, though some electricity was supplied by a wind generator. It was, he says, Spartan and functional, but they managed to make it cosy, eventually – not least because of the amount of polar clothing it had to contain. Its advantage was that it was only 6km (3.5 miles) from the emperor rookery on the sea ice at Auster.

To get to the rookery, the couple travelled by quads – four-wheel-drive bikes, which had Global Positioning System units mounted on the handlebars to aid

*Overleaf*: In a blizzard of up to 90kph (55mph), it was a battle to keep the camera steady. But the prize was footage of unique behaviour: penguins taking turns to stand on the more sheltered side of their huddle.

At the end of May, there were thousands of emperor penguins on the ice. The unsuccessful breeders would soon depart, but the males who had just begun to incubate their eggs now faced the Antarctic winter.

navigation in white-outs. Once there, they dragged the camera gear around on a home-made sledge fabricated from a pair of skis.

Emperor penguins are among the most remarkable and, frankly, appealing birds in existence – remarkable for their powers of endurance, appealing because of their winning ways, notably as parents and neighbours. At more than a metre tall, emperors are the largest penguin species. They live for more than 40 years, dive deeper than any other bird – up to 565m (1,850 feet) – and can stay under water for up to 22 minutes. There are almost 200,000 pairs, breeding on frozen ice, in 35 colonies scattered around Antarctica. They are the only species that breeds in the Antarctic winter – that is, during virtually the worst weather imaginable.

Lacking nesting materials, emperors incubate single eggs on their feet, covering them with their tummies. Males do the incubation, but since they cannot feed during incubation and hatching, they end up fasting for up to four months. For much of this time, the female is away, replenishing her reserves. She returns to help brood the chick, which can die within minutes if dropped on the ice.

The team drove for hundreds of kilometres before spotting a jade iceberg. This ice may have formed when the underside of the ice shelf melted and refroze, incorporating some green algae.

At the end of the filming day, when the team slowly retreated from the colony, a procession of curious emperors would follow the camera sledge (home-made from a pair of skis) for a few hundred metres.

'Provided we could endure the cold and isolation – which actually was wonderful – we had the greatest asset any wildlife camera crew can hope for – time, uninterrupted, to film in.'

The males survive their four-month fast not only through special adaptations that enable them to retain their body heat but also by a unique exercise in mass altruism known as 'huddling'. The emperor is the only non-territorial species of penguin, and it preserves its heat by packing together – sometimes as many as ten birds per square metre. The huddle has been described as 'an extraordinary act of cooperation in the face of a common hardship', with the birds actually changing position to share out the warmer spots. Hence the remarkable dissolving and re-forming patterns of the huddle – the 'huddle dynamics' – as the birds shuffle around, eggs on feet, moving to the lee of the huddle, where the wind is least cutting.

Fairley went with a script, or 'shopping list', of key sequences that were required for the series – the courtship of the emperors, the egg transfer from female to male, the huddling and hatching – but 'the script often plays out differently, because the only one not involved in its writing is the animal.' The emperors sometimes failed to live up to their caring image: for example, one sequence in particular involves chicks being violently fought over, because there

*Right*: It would sometimes take the team a couple of hours to clear the night's snowfall from their quads, which were used to travel the 6km from their hut, a refurbished 1950s freight container, to the rookery.

*Below*: They waited many nights for the conditions to capture the aurora above the huddle: no cloud, limited moonlight and an intense aurora in the right direction – the camera couldn't be moved once running time-lapse shots.

were fewer eggs than breeders, while another shows groups of healthy chicks being abandoned during a blizzard.

Blizzards tested even the emperors but presented a different dilemma to the film-makers. Penguin huddling in the wildest of winter conditions was high on the shopping list – but how do you film in a blizzard? And how safe is it to do so? 'We wanted images in the worst weather that the birds have to endure, to illustrate the behaviour and adaptation needed to survive,' Fairley explains. 'But it's a fine line between a blizzard with enough visual drama to work on film and one with too much real-life drama that it becomes a safety issue to work in. All of the blizzards were a battle. At times they were just plain frightening.'

But it was one of the first they filmed, in June 2005 – the depth of the Antarctic winter – that was the most formative. 'It was heart-breaking. To get to the rookery, we'd had to blunder along, half crouched in the wind gusts, partially blinded, our goggles filled with fine, driven snow and our eyelashes frozen together. Our hoods provided just enough tunnel vision to flounder along and find the birds. There were about 10,000 of them huddled in this whirling maelstrom of driving snow. The visual drama of the image was extraordinary. The birds were packed tight together like bricks, their communal bulk and low profile making them impervious to the storm. Great waves of driven snow were shooting over them. The power in the image was incredible – it was hard to believe that any animal could live in this and carry its precious egg throughout.'

'It's a fine line between a blizzard with enough visual drama to work on film and one with too much real-life drama ... All of the blizzards were a battle. At times they were just plain frightening.'

'It was the shot we had been waiting and hoping for, and so I framed up, trying to keep my head steady in the wind, buttoned on – and nothing happened. The wind was screaming so loud I couldn't hear whether the camera was turning, and so I tried again – and nothing. The magazine had jammed. There was no way we could change mags out there and no time to go back to the hut and then try again – the twilight wouldn't last long enough. I could have cried – if my eyes hadn't already been near frozen.'

For Fairley, this was the one that got away. Later attempts to film blizzards were more successful – 'we learnt a lot from that experience' – but they 'never seemed to have that same visual drama as the one we missed'.

The pair left Antarctica in November 2005, at the start of their second summer there, with images – and memories – of howling gales and zero visibility, of long periods spent watching, waiting and shivering, of startling displays of light, from the aurora australis to the brief appearances and final disappearance of the sun.

'It was an extraordinary experience to be living remote and independent in the world's greatest wilderness,' says Fairley. 'The winter light has a unique magical characteristic – a faint soft twilight. It's like nothing else. And provided we could endure the cold and isolation – which actually was wonderful – we had the greatest asset any wildlife camera crew can hope for – time, uninterrupted, to film in. We had the time, and we acquired the patience.'

# the lost world
## flying over the highest waterfall on earth

CARACAS
● Angel Falls
VENEZUELA

When the adventurers in Sir Arthur Conan Doyle's *The Lost World* set about discovering the mysterious plateaux in South America that might conceal a land of dinosaurs, they approached by boat and foot from Brazil to the south. When *Planet Earth* sent a team in, they were a touch more pragmatic – they flew in by Cessna from the Venezuelan capital of Caracas to the north. But though one trip was fiction and the other fact, the boundaries between the two often, and worryingly, seemed to blur.

Filming the *Planet Earth* programme on fresh waters meant travelling to the semi-mythical table-land in the southeast of Venezuela, and there were times during the trip, says producer Mark Brownlow, when it felt as though they were stuck in a remake of a 1950s Hollywood dinosaur movie. Sometimes it was funny – just about. Sometimes it wasn't. A lot of the time it was astonishingly, if predictably, wet.

The first *Lost World* moment came in the middle of the night at Caracas airport when they met their local fixer, who took one look at the 11 cases of gear they had brought. 'Do you realize where we're going?' he asked. 'We're going to the end of the world – in a small plane. You're never going to fit those in.'

Later came the encounter with the helicopter pilot who announced that he had 'only crashed five times' and that his machine, a mere 30 years old, was having a problem with its revs. 'You do raise an eyebrow when you see people filing down the heads of the earthing plugs,' adds Brownlow.

Finally, there was the place itself – hazardous enough without the lashings of mythology. The aim was to capture the drama of the rain and mist on the summit of Mount Kukenon, film the sandstone towers eroded into fantastical shapes on the mountaintops by the incessant blasting of the rain and, finally, secure footage of Angel Falls – at 979m (3,200 feet), the world's highest waterfall.

Kukenon and Mount Roraima have both been cited as the inspiration for Conan Doyle's novel. Brownlow took the book with him and says the exact location isn't clear. The two huge table mountains – known to local Pemon Indians as *tepuis* – glower at each other across the jungle in the Gran Sabana of southeast Venezuela. Local myth has it that the rocks 'call people to their deaths', and since Brownlow and cameraman Richard Burton were staying in an Indian village while attempting to film aerials of Angel Falls, the mood was catching – they were both 'a bit jumpy'.

To film the falls, the helicopter had to overfly another huge *tepui*, Mount Auyan (or Devil's mountain), and drop down into Devil's Canyon – risky because of treacherous mountaintop winds and the threat of white-out from fast-moving clouds. 'It's only a 20-minute helicopter ride, but for that 20 minutes you've got to cross your fingers and hope that the clouds aren't going to follow in. If they do, there's a good chance you're going to join the wreckage you can see strewn along the top of the *tepui*,' says Brownlow. They made it to the falls at the fourth attempt, after a 45-second white-out that rates as possibly the most alarming moment of the trip.

The footage of Angel Falls, and the 'gargoyle-esque' sandstone towers of the

*Above*: The clouds parted for just 45 minutes, enabling the team to secure one of the most dramatic sequences in the whole series – when the ground dropped away by almost 1km as they flew over the edge of Angel Falls.

*Left*: Travelling light was not an option – the team had 11 cases of filming gear. Much of it was specialist equipment: a heligimbal to reduce wobble in aerial shots and kit for the time-lapse photography of storms.

To capture the drama of developing storms, the team and all the kit were flown to the top of a *tepui*, or table mountain. The extremely wet, sheer-sided plateau rose 1,000m above the surrounding forest.

*tepuis*, is all the more impressive because it was filmed using 'heligimbal' anti-wobble technology (see p9) – part of the reason for the 11 cases of gear, since the heligimbal had to be brought in from the US and mounted onto the helicopter. For the footage on top of Kukenon – about 1,000m (3,300 feet) above the surrounding jungle – smooth, arcing shots were needed to give a sense of space and depth. This meant airlifting more specialist equipment – jibs, cables, camera 'dollies' and time-lapse cameras to film developing storms – onto the plateau and camping there for ten days while the shots were painstakingly pieced together.

Camping on Kukenon is not recommended. 'It rained incessantly. The tents were leaking, the sleeping bags were wet, the camp looked like a bath. We always had wet feet. The top is bog, and so when you walk through looking for a good position to rig all the kit up, you never know whether you're going to sink a foot or six feet. Get it wrong, and you're absolutely soaked. We wore oilskins and waterproofs, but the water gets through eventually, whatever you do.'

In the end, it was hard to separate fact from fiction and either from myth. Though Kukenon requires special permission to visit – most foreigners make for Roraima – a young German tourist, drawn to the area by its *Lost World* reputation, went missing on the mountain recently and was never found. 'Kukenon is a pretty spooky place. When the clouds come in, they can stay for days. The advice is just to stay put – it's all too easy to fall off the edge or down a crevasse or drown in a bog. It's known as Suicide Mountain – the Indians used to go up there to get lost.

'You can appreciate why Conan Doyle chose these mountains for his book. They rise up out of the lush, green jungle, and there doesn't appear to be any route up. The land on top is completely cut off from the world below. We got some really powerful time-lapse footage of clouds rolling in.

'It took us four goes to get to Angel Falls – it's almost permanently obscured by cloud – but at last, there was a 45-minute window of sunshine, and we grabbed our chance. The gorge leading up to it was perhaps the most primeval and prehistoric world I've ever witnessed. Richard [Burton, cameraman] summed it up when he said you could just imagine the pterodactyls sweeping past. The moment we flew over the edge was absolutely breathtaking.'

Formed by the incessant blasting of rain, the eroded sandstone towers on top of the *tepuis* added to the other-worldly atmosphere. To convey this, the team filmed from the air, using 'heligimbal' technology.

# walking on ice
## up close with polar bears

SPITSBERGEN

Kong Karls Land

NORWAY

There was a special poignancy about filming polar bears for *Planet Earth*. The survival of many large animals is threatened by habitat loss, but polar bears are peculiarly vulnerable. There are perhaps 25,000 of them left in the world. Their lives are entirely dependent on the formation of the Arctic sea ice, yet this ice is imperilled by a force that humans have let loose but seem incapable of halting – climate change. In a decade or two, the Arctic may be ice-free in summer – a state that could presage the complete disappearance of the polar bear's habitat.

Should this happen, the sequences filmed by Doug Allan and field assistant Jason Roberts in the northern spring of 2005 may take on extra significance. They encompass the crucial early stages in a cub's life – the 12 days or so between its emergence from the den and its move out onto the sea ice, where it will spend most of its adult life. Allan and Roberts caught virtually every moment on camera – almost certainly the first time anyone has done this.

To film the emergence of the cubs, the pair travelled to a place reputed to be, in Allan's words, 'bear nirvana' – the Bogon Valley in Kong Karls Land, a group of islands east of Spitsbergen, between the Barents Sea and the Arctic Ocean. The islands are governed by Norway, which protects them fiercely from the outside world – this was the first film crew ever allowed there and, says Allan, the first spring visitors of any description for a quarter of a century. 'The BBC had been trying for 25 years to get permission to go to Kong Karls Land, and for *Planet Earth*, it was finally granted,' says Allan.

There was one condition, however – to avoid disturbance, no snow machines were to be used. This might have put the crew at a serious disadvantage. Polar bears are dangerous animals – and if they look hostile, snow machines are a good means of scaring them away or, if that fails, of making an escape. In fact, the ban made for a richer filming experience.

'With a snow machine, driving five or ten miles [16km] is nothing,' says Allan. 'When you're walking, it's quite different. You become super-aware of the sounds, the temperature, the winds, the snow conditions. We were the only people within 150 miles [240km]. We were doing everything on foot and in the same time frame as the animals ... It felt the closest to being a polar bear it's possible to be.'

The result was some wonderfully intimate imagery of eight-week-old cubs venturing outside their den for the first time, and some disconcerting encounters with adults. Despite its reputation as bear nirvana – on the last visit by scientists, in 1980, more than 22 dens were reported in the area – the valley seemed underpopulated when the team arrived. They found only four dens in total, three of which were unworkable for filming – too high on very steep snow slopes. The fourth, however, was 'perfect': the crew dug their hide and began the wait.

The hide, or 'blind' – a hole with a snow wall round it – was about 80m (260 feet) down the slope from the den. 'Sometimes all you see for a couple of days is just the bear's nose sticking out of the snow,' says Allan. 'The bear may appear one day, and if the conditions aren't right, it may be a week before it comes out again.' In this case, however, the mother saw the hide, marched down the slope

*Opposite top*: The female bear finally emerged from her maternity den, where she had spent the past few months. She sat on the snow slopes, taking her time to assess conditions before bringing out her tiny cubs.

*Opposite bottom left*: A small cabin, known as Bear Corner, was home to Allan and Roberts for two months. At night, polar bears would sniff around it, scratch on the walls and peer through the windows.

*Opposite bottom right*: The team dug a hide, or snow blind, about 80m from the entrance to the bears' den just beneath the cliffs – but Allan had to stop filming when the mother bear came over for a much closer look.

The two eight-week-old cubs emerged onto the snow slopes only once or twice a day, staying out for just 20-30 minutes at a time before their mother squeezed back into the den, head first, and they disappeared in after her.

towards it, paused about 30m (100 feet) away, then kept on coming. When she got to about 10m (30 feet), Allan, who had been filming up to this point, decided on discretion rather than valour.

'Ten metres is getting too close for comfort. If a bear decides to attack, it can cover that distance in seconds. I decided I'd got enough close-ups and it was time to get it to go. So I pulled the explosive flare gun from where it was hanging on the ice wall of the blind and fired a round to frighten the bear.'

The mother saw the hide, marched down the slope towards it ... then kept on coming ... 'Ten metres is getting too close for comfort. If a bear decides to attack, it can cover that distance in seconds.'

The bear retreated but remained curious. So did the other local bears. There were several similar encounters during filming. At night, drawn by the smell of cooking food and other human aromas, the bears would often sniff round the men's tiny (three metres square) cabin, scratching on the walls, peering through the windows. Some bears could be scared away by the men simply opening the door and waving at them, others were more persistent. So Allan and Roberts rigged up tripwire fences attached to explosive flares outside the cabin, around the blind and around the snow-pits they dug near the den, where they stashed their filming gear.

'The mum and cubs knew we were there, but they were quite accepting of our presence,' says Allan. 'We watched them for 12 days, until they left the slopes to go onto the sea ice. Those were long days. The cubs came out for maybe only 20 minutes or half an hour, maybe twice a day, sometimes only once. Occasionally the mother would sit at the edge of the den for an hour or more but then decide it was too cold for the cubs.'

'The average temperature was -20°C (-4°F) and the average windspeed about

10–15 knots, and so the wind-chill factor was 50–60°C below [-75°F]. You have to keep moving to stay warm, and so all the time we were in the blind, we were shuffling around, doing little jumps and walking on the spot. We worked out that, though we were doing a five-mile round-trip to the hide, we were probably walking an extra ten miles [16km] inside it simply to keep the cold at bay.

'The very last day we were there, the mother gave me the final shots I needed. She was zig-zagging across the sea ice, hunting, followed by her cubs. It was very touching. When you go for drama and emotion in filming animals, you're often drawn to a predation sequence, but this was just lovely, intimate stuff. It's scary to think the species could be facing extinction in 30 years. If you take away the sea ice from polar bears, it's like taking away the forest from the tigers. It's as simple as that. Some of the sequences we were filming, you'd never be able to get again.'

At sunset, after each day's filming, the team made the long trek back to the cabin on foot – snow machines were banned to avoid disturbing the environment.

# beauty
# and drama

The threats were real – from angry whales and predatory mountain lions to deadly snakes and notorious bandits – but the rewards were great. What they observed and filmed 'was spectacular ... we had never seen anything like it.'

*Opposite*: On camera – the unpredictably violent walrus. *Left to right*: Gliding tree frog – endless nights sneaking up on amphibians. Local help – finding the best bird of paradise shows in Papua New Guinea. Spot the snow leopard – the film-makers' Holy Grail.

# frog heaven
## seeking a jewel in the forest

CENTRAL AMERICA

COSTA RICA • Liverpool

**The team tracked down the rare lemur leaf frog – not much bigger than the end of your finger – to its last natural breeding population at a pool high in the mountains of Costa Rica.**

It's a sobering experience to be present at the extinction of a species, but when you're filming frogs – particularly rare frogs – the chances of it happening these days are high. Amphibians are suffering a worldwide population crash: they're reliant on a habitat that's in increasingly short supply – areas with water suitable for breeding – and they appear peculiarly vulnerable to pollution and disease. Factor in the pressures from land clearance, and it's not surprising that a frog may almost literally be here one day, gone the next.

The lemur leaf frog is a tiny green jewel of a creature not much bigger than the end of your finger. It has skinny legs and delicate splayed hands and feet that look a little like the knobbly bits of a gaming jack. Natural breeding populations are known – or it might be more correct to say were known – at only one place in Costa Rica, which is where, in 2004, a *Planet Earth* team headed.

Costa Rica is one of the world's hotspots for frogs, which means it is also a hotspot for frog disappearances. Of the world's 5,000-odd known species of amphibian, the Americas are home to 53 per cent, and nearly two out of five of these are threatened with extinction. Montane Costa Rica is one of five areas in the New World where the threat is said to be greatest.

The *Planet Earth* team didn't go to Costa Rica to film extinctions, however. They went to film frogs. The idea was partly to demonstrate how sound, in the dense setting of the forest, is a vital mode of communication – and frogs make a

**All night, every night, for almost a month, they endured uncomfortable heat and biting insects, while waiting for the frogs to call. On the last night, they were rewarded with the sound of a wild lemur leaf frog.**

lot of sound. They also wanted to capture some of the great frog gatherings, when hundreds of individuals meet, fight, mate, lay their eggs.

Filming frogs has its idiosyncrasies. It helps if it rains, for example – so the team 'specifically picked the wettest time of the year to visit Costa Rica', says assistant producer Tom Hugh-Jones. In fact, it didn't rain once in the month they were there.

Then, since frogs are nocturnal, film-makers have to be the same: getting up at 4pm, eating dinner, working through the night, going to bed about 7am. Frogs don't much like lights, either. 'We spent whole nights sneaking up on frogs with the camera in the dark, waiting till they got used to it and then slowly turning the light on. But as soon as we did, they stopped calling. It was almost as though they were doing it to annoy us.'

Among the chief targets of the filming trip was the mass mating of the gliding tree frog, when hundreds gather under the light of the full moon. Andrew Gray, curator of herpetology at the Manchester Museum, had told the team how spectacular this was: after several nights of waiting, amid growing doubts that the frogs were in the mood, his forecast proved correct.

Finally, there was the herpetologists' Holy Grail – a pool up in the mountains near the east coast town of Liverpool. The scientists advising the team hadn't been there for a year, but it was known not only as a place of rich amphibian diversity but also as the only surviving home of the lemur leaf frog. On the last night of the trip, they set off into the mountains.

Says Hugh-Jones: 'There were eight of us in two cars, and it was a difficult place to get to. At the last town before we set off into the mountains, the bridge had gone. So we had to walk through the river to find a route across, and then drive the cars through it. There were landslides in the mountains, and we got to a point where the road was built into the mountainside and crumbling away. It was only just wide enough for a car. One vehicle with the local guys turned back – they said it was too dangerous.

'We got to the site just before dusk to find it had been fenced off. The landowner wasn't there, but a young man had been left to look after it, and he refused to let us in, saying he'd lose his job. By this point, we'd been through all these ordeals ... I got on my knees and begged him to let us in. Eventually, he got so tired of me that he agreed, but he said we had to be out before it got light.

'The site was near a pool further up in the mountains. There must have been thousands of frogs there, at least 12 different species. It was unbelievable. There were so many of them that they were quite relaxed. We filmed there all night – filmed them mating and egg-laying. We heard the lemur leaf frog calling and managed to film it, too. That was special for us, because it was so rare and charismatic.'

For the scientists who knew the site, however, there were mixed emotions. Squatters had moved in. The previous year, the pool had been surrounded by 50m (160 feet) of forest; now the trees had gone, and the pool was exposed to the cleared land. The filming took place higher up in what remained of the forest. Such was the speed of clearance, they told Hugh-Jones, that in another year, all the local forest would be gone, and, almost certainly, the frogs with it. 'It was a very special place,' says Hugh-Jones, 'but it wasn't going to be around for much longer.'

The future of species such as the lemur leaf frog may lie, for the foreseeable future, in breeding them elsewhere. The images taken that night may represent a small but tragic epitaph – the last glimpse, in Costa Rica at least, of a genuinely wild lemur leaf frog.

## a touch of venom

**When Tom Hugh-Jones was filming in Ecuador in 2004, a guide picked up a poison arrow frog – and passed it to him. Among the world's deadliest creatures, poison arrow frogs produce toxins from glands on their skin. Two hours later, Hugh-Jones wiped his brow with his hand. Within minutes his eye began to sting, then it turned blood red – which is how it stayed for weeks. Says Hugh-Jones, 'It still goes red now when I'm tired.'**

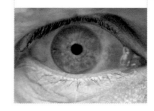

# the perfect predators
## stalked by mountain lions

**CHILE**

Torres del Paine
**PUNTA ARENAS**

What do you do when you're being stalked by a mountain lion? To be precise, by a family of mountain lions? There are a few tips – don't flinch, don't cut and run, stare straight back if you're being eyed up – but beyond that, you have to pretty much make it up as you go along. See how things develop.

It happened to cameraman Jeff Turner and producer Jonny Keeling when they were filming pumas in Torres del Paine National Park in Chile for the programme on mountains. For Keeling, it was an object lesson in, well, impetuosity.

'Sometimes you just rush in ... You think, great, we've got some cats here, let's do some filming before they vanish. And to be fair, there's no way of getting nice shots unless you're very close to animals. You've just got to be sensible.'

At Torres del Paine, a degree of impetuosity was excusable. The mountain lions here, known as pumas, are largely unstudied and reputedly reclusive. The crew bumped into a Chilean ornithologist who had led bird tours in the park for 15 years – and had never seen a puma. They also had the example of Hugh Miles,

'I think the mother must have spotted us ... she was staring at us with this completely fixed stare – glaring at us really. And then she got right down on her belly, the way cats do when they are stalking.'

*Opposite top*: Pumas are solitary except for females raising cubs – usually in litters of two or three. On their second trip, the team was very lucky to find a mother with four cubs, but keeping sight of them proved a challenge.

*Opposite bottom*: They watched the behaviour of the llama-like guanaco for clues to the pumas' whereabouts. Guanacos are regularly hunted by pumas (along with smaller prey such as hares) and are very alert to the cats' presence.

the veteran film-maker, whose programme on the pumas of Torres del Paine had won several awards in the late 1990s. But the film had been based on Miles habituating a female puma to his presence, and it had taken him more than two years to make – Keeling's team had six weeks. Miles came out to advise, but on the way over warned: 'There's a pretty good chance we'll come back with nothing.' Keeling recalls: 'He'd been telling me that all along, but he reminded me again on the plane. It gave me an awful sinking feeling in my stomach.'

In the event, he needn't have worried. Another team member was Robin Pratt, a Welsh livestock farmer specializing in guanacos – the wild llamas that are among the main prey of mountain lions. On the first day, Pratt left camp not long after 3am in his vehicle and within 10 minutes was calling on his walkie-talkie in a state of high excitement. He'd seen not one but four pumas – a mother and her three almost grown-up cubs.

Keeling says: 'We jumped in the vehicle and raced out. We found the mother and three cubs. But then they vanished. It was a big open area, and there were four or five of us in two vehicles with walkie-talkies, but they just melted away. I was quite angry – we'd had this great bit of luck and I thought we'd blown it. But that's why they're such great hunters, I suppose.'

More evidence of the pumas' hunting skills was to follow quite quickly. After two hours or so, Keeling spotted the four cats next to a lake about a kilometre away. 'I contacted Jeff through the walkie-talkie, we got the camera gear and quietly approached them. We got to some bushes within 300m [980 feet], and they still hadn't heard us or smelt us. They were playing on a shingle bank next to the lake.

'Jeff took the camera out of the backpack. I think the mother must have spotted us. We were looking down getting the camera gear ready, and I looked up, and she was staring at us with this completely fixed stare – glaring at us really. And then she got right down on her belly, the way cats do when they are stalking. Her head was locked onto us, her body was turned slightly sideways and she started creeping forward to the nearest bush. A couple of her cubs copied her and started crawling towards this bush.

'Jeff is a very seasoned photographer, who's worked a lot with dangerous animals ... He turned to me, looked really frightened and said she was stalking us. I had chucked some pepper spray into my pack – for use as a last resort to spray in the face of an animal that attacks. He told me to grab the spray and stand next to him – it was important that we were together. I was very happy to do that.'

Cameraman Jeff Turner was stalked on several occasions, particularly by the cubs, as here (to the right of this tripod). They may just have been curious, but attacks on people are not unknown. The crew carried pepper spray just in case.

As the pair waited, the pumas performed another of their disappearing acts. Minutes later, they turned up in bushes about 20m (65 feet) away – Keeling spotted a flicking ear and the same level gaze. 'We were surrounded by trees and bushes, but there were areas of open space, too, which she had covered without us seeing her. She went down into the grass. One of the cubs came out and walked up to within about 15m [50 feet] of us, sat down and stared at us. Then its eyes started closing, and it went to sleep in the sun. I couldn't believe it.'

Would the pumas have attacked? Mountain lions do kill people – a fisherman was mauled to death not long ago in Torres del Paine itself. When the pair approached the pumas, they were torn, in Keeling's words, between 'not wanting to scare them away and not wanting them to attack us'. He adds: 'You're surrounded by bushes, and you're being stalked by several large predators whose whole lives – whose whole evolutionary history – is geared up to hunting.'

'The important thing with a puma is to keep eye contact. It's the same with other lions. If they think you're looking the other way, if you turn your back or start running away, then that can elicit an attack. With bears and gorillas, you look away, you're humble. From the point we stared back at her, when she was 20m [65 feet] away, we were probably safe – we had stared her out.'

In the event, the puma family came to treat the film crew as part of the background – to the extent that the crew would often sleep out rough in their sleeping bags near a fresh kill, filming the pumas feeding on the carcass at night and moving off with them at first light next morning. 'I didn't sleep very deeply ... it was the most fantastic feeling to be so close and for them, if not to accept us fully, then at least not to kill us. But at the beginning, you never quite know.'

Filming was often in low light, at dawn and dusk, when the pumas were most active. As the cats grew more relaxed in the crew's presence, the team could sleep out in the open nearby to begin filming at first light.

# the dark side of light
## struck by a deadly serpent

SARAWAK
● Bako National Park
**BORNEO**

*Opposite:* **First the team had to find a colugo, usually well camouflaged on a tree trunk. Then they had to illuminate the forest to film it gliding between trees in search of food – mainly leaves, shoots, buds or flowers.**

*Above:* **The venomous Wagler's pit viper belongs to a group of snakes that includes the rattlesnake and water moccasin. It locates prey – and perhaps hot filming lights – with specialist heat-receptor pits.**

Driving the wrong way down a two-lane highway in Sarawak looking for an ambulance is a situation you'd hope to avoid if you're making a film about colugos. But if you've just been bitten by one of the world's more poisonous snakes, you don't have much choice.

Researcher Jeff Wilson admits there was a touch of farce about the night a pit viper bit him in the forests of Bako. The fact that it was 2.30am, the nearest hospital was a boat journey away and the boats were beached because it was low tide just added to the entertainment.

Colugos are better known as flying lemurs, though, as Wilson explains, they're not lemurs and they don't fly – they glide with the aid of large, webbed 'wings'. Because they're nocturnal and well camouflaged, they have seldom been filmed.

Wilson, cameraman Gordon Buchanan and a Japanese film crew went to Bako National Park with lighting equipment designed to floodlight the forest and capture the colugos gliding. HMI lights – short for Hydrargyrum Medium Arc-length Iodide – are highly efficient and produce a sunlight-like effect. But they had to be hauled round the forest at speed and gave off a lot of heat.

'Gordon had got his camera into filming position, and we were moving the lights,' says Wilson. 'We were just about to nail probably the best shot we'd ever got. Without looking, I put my hand on the light stand and felt the strike. I knew immediately it was a snake. When I saw it was a pit viper, I knew I was in a bit of trouble.' Wagler's pit viper, which has specialist heat-receptor 'pits' designed to sense prey – or possibly camera lights – is the only venomous snake in Bako.

Wilson knew that panicking would help pump the venom round his body. So he tried to stay still, feeling his arm turn numb, his back and neck muscles stiffening, while the others set about getting him to hospital – 40km (25 miles) away, by sea and road. The boatman refused to budge at first, because the boat was stranded by the tide. Lights were rigged up to avoid hitting logs, but the engine kept cutting out when the boat ran into seaweed. On the mainland, the promised ambulance failed to materialize, and so they drove on the wrong side of the road, dodging oncoming traffic, to make sure the ambulance didn't pass them. Hospital staff then queried how they could be sure it was a pit viper. 'We got a bit angry at that point and said it was our job to know,' says Wilson.

Experience of snake bites had taught Wilson that there's usually a 'crunch point' about an hour later 'where things could go downhill quickly or could be OK'. On the boat, he'd concluded he was probably going to survive – or at worst, lose a finger. 'For about 48 hours, I felt as though I'd had a dead-arm competition with my brother. Other than that it was fine.'

There was a postscript, however. One reason he survived was probably the location of the bite. 'The worrying thing was that I had been carrying the light on my shoulder for several minutes – with the snake next to my ear ... If it had bitten my neck, it could have been Goodnight Irene.'

# the bandit king
## tracking otters in dangerous territory

INDIA

KARNATAKA
Cauvery River ●

Filming smooth-coated otters on the banks of the Cauvery River in India was never going to be easy. The animals are not only endangered: they're clever, wary and prone to disappear at the slightest sign of human presence. But it got a lot more difficult when the film crew found itself embroiled in a war between the state and a local bandit king.

Cameraman Charlie Hamilton-James and assistant producer Kathryn Jeffs went out to Karnataka state, not far from Bangalore, in 2004. They spent five days scanning the riverbanks for signs of the animals, hoping to track some back to their holt. On the sixth day, they thought they had struck lucky.

'We were getting up at 5am, before it got light, and setting out along the riverbank with our heavy kit – tripods, cameras, binoculars,' explains Jeffs. 'We were keeping in contact by radio, with Charlie in complete camouflage gear – gloves, hat, face-shield, pants, shirt. When the sun came up, it was incredibly hot – poor Charlie was streaming with sweat. He was so well camouflaged that I could be staring at the riverbank with my binoculars and unable to see him. I'd ask him where he was over the radio, and he'd turn his hat inside out to show a little fluorescent sign. I'd find that I'd been looking right at him.'

'We got some shots of the otters fishing, but we were trying to find their holt. We were working with trackers from a local tribe, and on day six they came to us very excited and said they thought they'd found it. So we hunkered down and waited, and pretty soon we heard the otters squealing and shrieking as they returned to the holt. Then it got hot and they settled down to sleep, and we needed to drink and eat because we'd been working since 5am. So we went back to camp.'

That same morning, however, Hamilton-James had made a macabre and

*Opposite*: Every morning, the team set out along the banks of the River Cauvery in southern India to look for signs of otters. What they weren't expecting to find was a human skull with a bullet hole in it.

*Below*: With the help of trackers from a local tribe, they managed to locate an otter holt and film the otters returning to it. But with the area becoming increasingly unsafe to work in, filming time was limited.

Despite the difficulties, they did get some footage of smooth-coated otters on the riverbanks. The animals were creatures of habit – local fishermen said they could set their watches by the otters' appearances.

unnerving discovery on the riverbank – a human skull with a tiny round hole in the temple. And on returning to the camp, there was more disconcerting news. The camp manager had been told from Delhi that a 'situation' had arisen.

The area they were working in, the film crew learnt, was the unofficial 'territory' of the bandit king Veerappan. Veerappan was a highly dangerous man, thought to have been responsible for the murders of more than 120 people and running some 6,000km² (2,300 square miles) of jungle in the Western and Eastern Ghats in southern India almost as though they were his own personal fiefdom. His jungle 'reign' had lasted for 30 years, taking in poaching, smuggling and kidnapping and numbering some famous Indians among his victims. In 2000, he took the film star Rajkumar hostage, and two years later, he kidnapped and killed a former Karnataka minister. In his heyday, in the early 1990s, he was the most notorious of all the Indian dacoits – better known even that the 'bandit queen' Phoolan Devi.

The 'situation' involved two of Veerappan's men, who had been serving life imprisonment for crimes in his service but had just had their sentence changed: they had been condemned to death. The word from Delhi was that Veerappan could be planning retaliation – by kidnapping the *Planet Earth* film crew and holding them hostage in return for his men's lives.

The crew's reaction, initially, was disbelief. 'I thought – this can't be true, this sort of thing doesn't happen,' says Jeffs. But it seemed that it did – a BBC natural history film crew had narrowly escaped kidnapping about two years before, the team was told. On that occasion an Indian film crew had been taken instead. And conversations with locals suggested that the threat posed by Veerappan was all too real. He had even killed one of his own daughters, Jeffs

First the film team was told it would be able to work only during the day and with an armed guard. Then ever-larger contingents of police began to arrive ...

In complete camouflage gear, cameraman Charlie Hamilton-James was barely visible on the riverbank. With the sun on him, it was very hot and uncomfortable, but he was able to get close to the otters.

learnt – he feared the infant's crying might give away his whereabouts to the authorities. 'It started to dawn on me that this was very serious,' she said.

The Indian authorities had no doubts at all. First the film team was told it would be able to work only during the day and with an armed guard. Then ever-larger contingents of police began to arrive: they warned the crew against going into the field and threatened to accompany them wherever they went. Finally police advised the team to leave. In such circumstances, filming was impossible: the shoot was abandoned and the crew departed. Days later there were reports of an 'attack' on the camp.

It was an affair with an ironic twist to it. In October 2004, after a manhunt that involved nearly 2,000 police from the two states of Karnataka and Tamil Nadu and was likened by some in India to the pursuit of Osama Bin Laden, Veerappan was caught and shot dead. The crew returned to finish their filming, but found the place much changed.

Veerappan – full name Koose Muniswamy Veerappan – began his criminal

career as an ivory poacher and is reputed to have killed his first elephant when he was only 14. From the late 1960s, when he started his gang, he is said to have killed more than 200 'tuskers': he was also involved in smuggling sandalwood.

'Veerappan had been shot dead when we came back,' recalls Jeffs. 'The place seemed different. There were poachers everywhere – during the day they would walk past us very blatantly with guns. They kept wrecking our chances of filming, scaring the animals away, shooting. There were a lot of bush fires. The otters themselves were much scarcer and warier. The local fishermen said they had been able to set their watches by the otters' appearances the previous year; on our second trip, they told us they hardly saw the animals at all.

'It seemed to me that Veerappan had had such a strong hold on the area that no-one could poach there without him knowing. And though he poached, it was mostly elephants and sandalwood, not smaller animals such as deer and otters. In the end, getting decent footage of the otters was really difficult – especially after Veerappan's death.'

Working from a raft on the river enabled the crew to get close to their subjects without the need for full camouflage – the otters seemed less wary of people on the water than those on land.

# night of a trillion bugs
## filming the world's biggest insect emergence

USA
Bloomington ●
INDIANA

It was touch and go whether *Planet Earth* would bother filming cicadas emerging from the ground in Bloomington, Indiana, in May 2004. Bloomington isn't the world's most exotic location and cicadas are 'just bugs'. In the event, it was a sequence that might have come straight from the *X Files*.

Producer Mark Linfield initially doubted whether the small, brown insects could provide enough of a spectacle. What turned it was not only the sheer scale of the phenomenon – with more than a trillion cicadas, it's said to be the largest known insect emergence – but also the extraordinary nature of the cicada life-cycle.

For 17 years the so-called periodical cicada of Brood X lives underground, sucking tree sap, before emerging as a brown, bug-like nymph, shedding its case almost immediately to pupate into an adult, flying, mating, laying eggs in trees – whereupon the larvae disappear underground, the two-week above-ground stage is over and the 17-year underground cycle begins again. Different cicadas have different periodicities – Brood X has the longest – and each is a prime number. Predators thus can't easily synchronize their life-cycles with the cicadas and gobble them all up as they emerge – which the scale of the emergence makes virtually impossible anyway.

Bloomington is the epicentre of Brood X's emergence, and since the cicadas are not far off the length of your little finger, it was definitely not a moment for the insect-averse.'The scale took me completely by surprise,' says Linfield. 'Each night, more bugs than you can imagine pour out of the ground. You think there can't be any left, and the next night the whole thing starts all over again. And it carries on, night after night, for six or seven days. You're ankle-deep in cicadas. You wonder why the ground isn't hollow, why it doesn't subside under you.'

The nymphs are extraordinarily single-minded. 'As far as you can see, there are seething bodies on the ground, and they're programmed to walk up anything vertical. So if you're standing still, they slowly march like zombies up your legs. They get to the top of your head, and then they'll split. The adult will come out and start hardening off in your hair. If you're really focused on filming, you somehow don't notice that you've got these things pupating all over you.'

Around you, meanwhile, are scenes from a kind of insect Armageddon – an overwhelming stench, carpets of dead cicadas, birds that have eaten so many they can't take off, squirrels and chipmunks 'like little balls rolling round and vomiting – they just can't squeeze in one more cicada'.

Many people in cicada-land won't go out at night during the emergence. Restaurants meanwhile serve cicada kebabs, cicada fritters. But what really struck Linfield was the deceptive ordinariness of cicada terrain. 'The highest densities are in back gardens, golf courses, university campuses – what the scientists call the suburban savannah. There were 16-year-olds there who, before May 2004, had maybe never seen a cicada. Yet the bugs had been living in the soil under their gardens for 17 years, and the kids never knew it. It really is like a horror film.'

The cicada nymphs crawl up anything vertical – including film crews – then, gripping the surface (hair will do) with their front claws, they pupate into winged adults (inset), wriggling free from their nymphal exoskeletons.

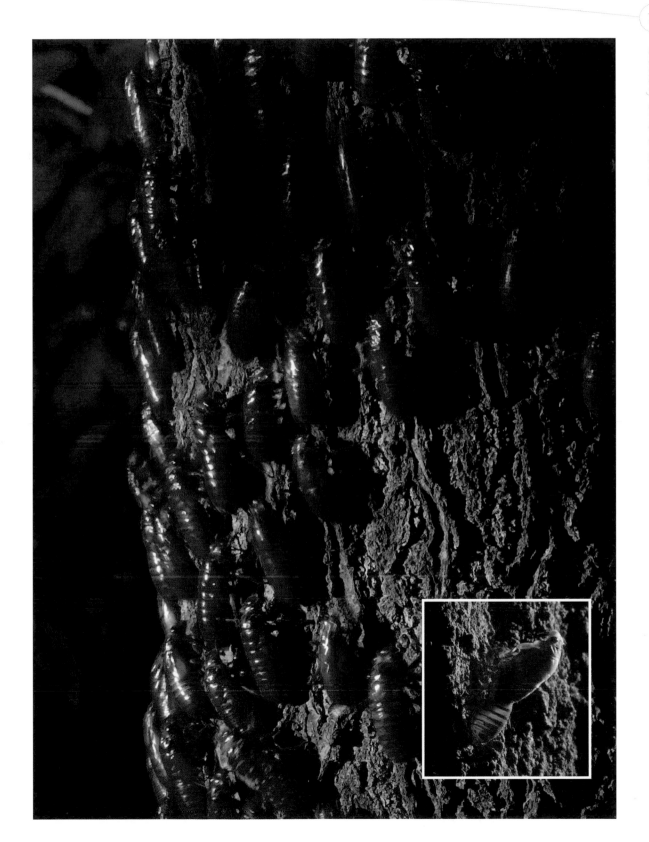

# band on the run
## chasing the ultimate hunt

BOTSWANA

Okavango Delta

**SOUTHERN AFRICA**

*The dogs simply flushed an impala out of the forest and ran it down ...*

*'The whole pack circled the lake and waited. We thought the impala was a goner.'*

*Overleaf:* **To capture the full drama of a wild dog hunt across the Okavango Delta's mosaic of habitats, some of the team took to the air.**

A happy ending is not something you associate with a hunt, particularly where African wild dogs are involved. The dogs pile into their prey and start to disembowel it while it's still alive. It's a gruesome spectacle – but it's one the makers of *Planet Earth*, albeit with some reservations, set out to film.

The long-distance drama of a pack-animal hunt, whether wolves or wild dogs, is seldom seen on television. 'There's very little actual hunting in previous films about the dogs,' says producer Mark Linfield. 'You see them setting out, then you see them on the kill. There's a lot of hand-waving about what goes on in the middle.'

Habitat is one reason. To film the sequence that appears in the opening programme of the series, a *Planet Earth* team spent two weeks in the Okavango Delta – a mosaic of grassland, swamp and scrubby mopane forest. Even if vehicles could drive in a straight line, the ground is littered with hazards, from mud and thorns to aardvark holes. And vehicles can't supply aerial shots.

The team used two reinforced, waterproofed four-wheel-drive Land Cruisers and a gyro-stabilized camera mounted on a helicopter – the heligimbal (see p9). 'It was a real pincer-movement attack,' says Linfield. They spent more than 50 hours flying but got the shot they wanted 'literally in the last five or ten minutes ... We had a predetermined budget, and we couldn't have gone on any longer.'

What Linfield and the team of cameramen – Michael Kelem in the helicopter, Martyn Colbeck and Michael Holding in the vehicles – hadn't reckoned with was the complexity of the hunt. 'The dogs most often hunt impala, and in a straight-line run, the impala are considerably faster than the dogs,' explains Linfield. 'What I don't think anybody had realised was how much the dogs were relying on the impala to make a mistake in order to catch them.'

Impala were frequently caught because they ran into the forest and blundered into trees – so the kill was made in 'pretty much the one place on the landscape that's invisible from a helicopter'. And it wasn't usually the front dog that brought down the impala; often the antelope would veer off, into the path of another dog.

'If you've got 10 or 12 dogs in the pack and they run at 50 impala and then go off in all directions, you have to commit to a single animal – particularly when you're concentrating on the long lens of the heligimbal, with its very narrow angle of view. From the helicopter, often a kilometre high, you can't see with the naked eye what the other dogs are doing. So a lot of the time we filmed the wrong dog.'

In the final sequence, however, the dogs simply flushed an impala out of the forest and ran it down, whereupon the antelope jumped into a small lake. Linfield adds: 'The whole pack circled the lake and waited. We thought the impala was a goner. But then one of the dogs made a kill in the forest, and the dogs by the lake pulled back to that. The impala got out of the lake, looked over its shoulder and took off. It looked as though it couldn't believe its luck.'

Though not a pull-down, it was certainly a striking finale. 'To be honest I was quite pleased, because a real hunting dog bring-down is quite a gory thing,' Linfield adds. 'I think we filmed something considerably nicer.'

*Above*: Wild dogs rarely return from a hunt without a kill – here, an impala.

*Top*: In a dry patch, between woods and swamp, the ground crew waited for a pack asleep under a tree.

*Left*: Cameraman Mike Holding radioed the aerial crew to say the dogs were about to go hunting.

# the ocean blues
## scolded by a sperm whale

**PORTUGAL**

● Azores

*ATLANTIC
OCEAN*

**M**aybe it was a lesson in sticking to what you are supposed to be doing. Or maybe it demonstrated the wisdom of not coming between a mother and her young. Whatever the explanation, cameraman Doug Anderson won't quickly forget the time he was given a good ticking off by a female sperm whale.

Anderson and assistant producer Penny Allen were filming for the open oceans programme off the Azores in the summer of 2005 when they spotted a medium-sized whale – about 8m (25 feet) long – and her calf. 'Though it was not really on our shooting list, we couldn't resist a look,' he says.

The pair climbed into an inflatable to approach the whales. About 30m (100 feet) away, Anderson slipped into the water with his camera. Looking down, his heart 'immediately started racing … [The mother] had not one but two babies – and they were tiny. One was with her about 15m (50 feet) below the surface, rubbing and suckling, the other was on the surface with what looked like a buoyancy problem.'

'Small whales can seem almost to forget how to dive for a minute or two before rediscovering the skill and shooting off. As soon as the baby spotted me, it came for a closer look. It was about the size of two big bathtubs and a perfect miniature of mum except for the foetal folds that had not quite come out of its belly.'

To Anderson's surprise, the calf came even closer and started 'to give me a good rubbing. Something no-one ever tells you is how hard whales are. Being bumped into gently by even a small whale is a bit like being hit by a fence post. But I was quite happy to put up with being pushed around for a bit.'

'I rolled a couple of shots close up and then dived down to get a shot of it against the surface.' By this time, he had lost sight of the mother whale and was enjoying the intimacy of the encounter. It proved short-lived.

'The baby came in for another rubbing session and caught one of its folds on the lens shade of the camera housing. It went mental. Sperm whales are incredibly tactile, and the feeling of metal on its skin must have been so different from what it expected that it shocked the little guy. He swam off at high speed, clicking like mad.

'I was gutted – for about three seconds – and then I was terrified. I looked down to switch the camera off, and I saw mum about 15 feet [4.5m] away looking straight at me with her jaw dropped. Her head seemed to take up my entire field of view. I could see the individual teeth, white and glistening, on her lower jaw.'

How does a sperm whale express its displeasure? 'There was a powerful zap of echolocation right in my chest. I don't know how long it lasted, but it felt like standing right next to the speaker at a rock concert. The stream of fast, loud clicks vibrated my sternum and seemed to joggle all my organs.'

Fearing an attack, he lifted the camera housing to protect himself and tried to calculate the distance back to the boat. But the whale had evidently had her say. 'When I looked down, she had closed her mouth and was falling away into the blue, rubbing and suckling with her babies … When I got back on the skiff, my hands were shaking. But I had probably never felt so safe in my entire life.'

**To film close-ups of sperm whales – the largest toothed whales, up to 18m (60 feet) long – it's often better to snorkel (with a small air bottle for safety) than to dive. Snorkelling is quieter, faster and doesn't risk disturbing the animals with bubbles.**

# giants of the shallow seas
## slapped by a one-ton whale calf

FIJI

● TONGA

SOUTH
PACIFIC

Cameraman Doug Allan was faced with a tricky choice after his wife Sue Flood was thumped by a hyperactive, six-metre (20-foot) humpback whale calf in the seas off Tonga. On the one hand Flood was clearly in some trouble, though still conscious and able to swim. On the other, about £15,000-worth of high-definition camera gear that she had just dropped was drifting down into the depths of the South Pacific. Which should he rescue – wife or camera?

The camera, of course. And he wasn't 'wildly impressed', says Flood, by her insistence on going to hospital. 'They managed to get me on the boat, which was about half an hour from shore, but as the visibility was great and the whales were very cooperative, Doug was desperate to stay out there. Eventually we did go back, and I went to the local hospital, which was a bit like a bus shelter and didn't have an X-ray machine. So they gave me a Nurofen.'

It's probably fair to say that Allan and Flood, both trained divers and photographers, have been battle-hardened by years of working together. The humpback footage, which appears in the shallow seas programme, is one of

many marine sequences they have filmed. As it happened, nothing was broken – Flood's leg was X-rayed later back home in Bristol – and in a few days, she was back in the water with the same humpback calf. 'She was very professional,' says Allan. But at the time of the collision, it felt a little more traumatic.

'A new-born whale calf weighs more than a ton, and so it was like being whacked with a sledge hammer,' Flood recalls. 'I was snorkelling and videoing Doug filming the calf and its mother, and I could see [the calf] getting closer and closer in my viewfinder. I don't think it was aggressive, just playful and boisterous and curious. It came up to me, rolling this way and that, and then it bumped into me. I think it gave itself a little bit of a fright and recoiled, and its tail hit my ankle and my thigh. I let go of the camera. The pain was so intense that I thought I'd broken my ankle.

'I got back in the water with the same calf three more times after that, and it did the same thing, rolling around and behaving boisterously, except that it didn't hit me or bump into me again. The other calves weren't doing that, and its mother was quiet and docile. We nicknamed it KC – Killer Calf. If it had hit my head, it would probably have broken my neck.'

'The odd thing was that I had just said to the skipper of the boat how lovely it was to be filming with animals that are so incredibly big yet so gentle. I've been in some tricky situations before – standing on the ice about eight metres (26 feet) from a polar bear, diving with leopard seals – but the thing that really frightened me was this friendly little whale calf.'

'Its tail hit my ankle and my thigh. I let go of the camera. The pain was so intense ...'

Like other baleen whales, the six-metre-long humpback calf – nicknamed KC, or Killer Calf – has no teeth and will grow up to eat nothing bigger than small fish, krill and plankton. It got its fearsome name from its unusually boisterous behaviour, which was almost disastrous for the team.

# the shape of paradise
## catching the birds' greatest shows

Tari
PAPUA NEW GUINEA

AUSTRALIA

If birds of paradise are among the world's most eye-catching creatures, the humans they live among can be pretty unusual, too. *Planet Earth* made three trips to Papua New Guinea to catch the birds' remarkable mating displays, returning with sequences never filmed before – including the intricate and acrobatic displays of the blue bird of paradise. But for those involved in the shoot, the action off-camera was almost as memorable.

There are more than 40 species of birds of paradise, many are threatened by hunting and habitat loss, and most are found in Papua New Guinea. To film the displays of four species – the blue, the superb, Lawes' parotia and the brown sicklebill – a team travelled to the Tari valley in the central highlands, the home of the Huli people. It was an illuminating experience.

'I was doing recces for the shoots, trying to find the best display sites to film at, and I would hook up with landowners who claimed to have the birds on their land,' says researcher Jeff Wilson. 'More often than not they had 100 or so bird-of-paradise feathers stuck in their hair. When the guy who is leading you to the birds has got half the local population on his head, it doesn't exactly fill you with confidence.'

It was a similar story when he was invited inside a Huli hut during a downpour and shown the owner's collection of arrows. 'They were all different shapes and sizes, and he explained that some were used for killing birds of paradise, some were for killing wild pigs and some – the very heavy ones – were for killing humans. They looked as though they were used quite frequently. They were all covered in dried blood.'

The Tari valley is a place in transition. Opened up to the outside world only since the 1950s, it's sufficiently remote for anthropologists to conduct fieldwork there, but it's increasingly a destination for upmarket tourism. The Huli, former cannibals who wear wigs made from human hair and paint their faces ochre, are known as the Huli 'wigmen'. They were welcoming to the BBC crew, but filming there, nevertheless, had some uncomfortable moments. A clan war broke out when one family's pigs ate a neighbour's sweet potatoes. Cameraman Paul Stewart had arrows fired at a hide – though not when he was in it. And the *Planet Earth* team found small things routinely going wrong.

A wooden road bridge used by the crew to access the hides kept losing its planks, for example – the wood would later turn up in a Huli hut extension. And the team's vehicle would regularly grind to a halt in an apparently purpose-built Huli quagmire.

'They would wait there every day pretending to repair [the ground], and when the car got stuck, they would all be at hand to pull it out. Everyone would be paid, and they would give you a wave as you left. But if you put your head back round the corner, they would be urinating or putting more water down. They were very enterprising.'

By comparison, the filming itself seemed almost humdrum – albeit gruelling. For

In their mixture of Western and traditional dress, the Huli people showed a great interest in any form of technology used by the crew. Local culture has been adapting to outside influences since the 1950s.

most species, 'the routine involved getting up at four in the morning, arriving at the hide at dawn before the bird woke and being there, still, with the camera, when it started displaying. It would display between dawn and 10am. Around the middle of the day, it would rest or go off to a feeding tree, and so we would come out of the hide, stretch our legs and have a bite to eat before going back inside. The bird would kick off [displaying] again at about 3.30pm and go on until five or six in the afternoon.' Cameraman Stewart 'spent 8 weeks in a hide, 14 hours a day, 7 days a week.'

Hunting, supposedly illegal, had made the birds extra-sensitive to human presence. To acclimatize them to filming, the team left fake cameras in the hides made out of tubing, tape and reflective material 'so the birds got used to something black and shiny pointing at them'. Even so, one of the birds, the brown sicklebill, could not be filmed within the time available.

The pace of change in Tari was visible even over the 18 months the *Planet Earth* team were visiting. Western T-shirts proliferated, Western-style rubbish – plastic bags, tin cans – seemed suddenly to arrive and the villagers gained new familiarity with digital cameras. But some changes, sadly, failed to stick.

The team enlisted the help of local landowners to locate the best display sites. The family living where the blue bird of paradise and parotia were filmed spanned four generations – despite a life expectancy of about 50 years.

The Hulis couldn't at first see what benefits having live birds of paradise on their land would bring them. The *Planet Earth* team talked to them about tourism and conservation, pointed to the money the villagers were being paid for filming. 'By the end of our first trip, we thought we had convinced them,' says Wilson.

On the second trip, however, he had a ten-day scramble to find new filming sites. 'Pretty much all the sites we filmed the first time had been destroyed. People had let their pigs roam in the forests where we had been filming. They had chopped down crucial trees where the birds were displaying. I never quite figured out from talking to them what the reason was – whether it was disputes between families over who got the money from the filming, and someone had decided to chop down the trees and stop the whole business, or whether they genuinely just did not realize what they were doing.'

Researcher Jeff Wilson tracked the movements of the blue bird of paradise through the forest to find its display site. The team's goal was to capture the acrobatic mating display, which had never been filmed before.

# the stuff of dreams
## searching for snow leopards

AFGHANISTAN

● Northwest
Frontier Province

PAKISTAN

If you were awarding points to wild animals based on mystique, snow leopards would come out near the top. It's not just that they're beautiful or even that they're endangered – though with as few as 3,500 left in the world, they're certainly that. It's as much to do with their elusiveness, the remoteness of their habitat, their solitary, high-altitude lifestyles. For makers of wildlife documentaries, the mystique is particularly powerful. Snow leopards are hard enough to find, let alone film.

Cameraman Mark Smith calls the snow leopard 'the Holy Grail of film-makers – the ultimate in elusiveness, rarity, invisibility'. As he discovered in the mountains of Pakistan during a heavy snowfall in the winter of 2005, you can be looking straight at a snow leopard with your naked eye and you won't see it. Then again, you're doing well if you're looking at a snow leopard at all.

*Planet Earth* features some of the most compelling sequences of snow leopard behaviour ever filmed – including a remarkable hunt-chase across a near-vertical mountain face amid the snowy massifs of northwest Pakistan. Nothing quite like it has been seen before. And no-one was more surprised than the *Planet Earth* production team.

Politically, they could hardly have arrived at a worse time ... the local police chief insisted on giving them an armed guard – who slept outside their door with an AK47 automatic rifle in his bed.

*Opposite top*: Camouflaged against the rocks by the cave entrance, the mother kept a look out for markor (mountain goats) that strayed her way. What the film crew didn't realize at first was that hidden inside the cave was a cub.

*Opposite bottom*: Cameraman Mark Smith and the team, including a number of local trackers, spent every day scanning the cliffs for snow leopards. After three weeks, they had seen nothing more than a few tracks in the snow.

Smith and researcher Jeff Wilson flew out to Pakistan in December 2004 with little hope of success. It was a kind of 'double-bluff', Smith says. 'Ostensibly we were going out to film markor – mountain goats, which are the main prey of snow leopards. We thought that by filming markor we might also get some snow leopard footage. But I don't think anybody really believed we would.' As a result, they went out 'quite frugally equipped – just a normal high-definition camera and a long lens'.

Their destination was an area close to the Afghan border, and politically, they could hardly have arrived at a worse time. Terrorists bombed an aid-organisation centre a few days later, there was talk of the BBC crew being targeted and the local police chief insisted on giving them an armed guard – who slept outside their door with an AK47 automatic rifle in his bed. The weather wasn't the best either – the worst snowfall for 25 years had closed passes, shut down scheduled flights and rendered the region inaccessible except by army helicopter. They had got in, but when they would be able to get out was anybody's guess.

Worse, they went to the wrong place – to begin with. Local people described to them a valley in the mountains where snow leopards had been regularly sighted.

**With a huge snowfall, the mountain passes were cut off and the team stranded. But they thus gained extra filming time in conditions that drove animals down from the peaks and made tracks easier to follow.**

The valley was narrow and accessible by track, but it was cut in two by a fast-flowing river that nobody could cross. The animals felt safe beyond the river and so did not flee when humans approached on the other bank. It all seemed too good to be true.

Christmas Eve 2004, three weeks into the six-week trip, found Smith busy filming wolves and markor on a ridge 16km (10 miles) away from the valley without having seen a single leopard – though some tracks had been spotted in the snow. Then a message came through to the team on the radio – a kill had been sighted.

'The kill was in the valley where everyone had been telling us we should have been filming. But if a snow leopard makes a kill, there's a good chance it will remain there for a couple of days, and so we sprinted back to camp, got together as much stuff as we could and got over to the valley by the evening.

By then, the leopard had gone. It turned out there had been a delay in getting the message to us.'

Obviously disappointed, Wilson went back to the first location, while Smith decided to stay in the valley 'on the off chance ... I was feeling very miserable the next morning. I had a really bad stomach bug, it was the day after Boxing Day, it was very grim and rainy and it was all going wrong. By this time, I'd become totally obsessed – every rock I saw I thought was a snow leopard. We were walking up the track in the valley, and I saw something that did look like a leopard. I looked at it through my binoculars, and it was. It was on a ridge, sitting like a sphinx looking down on us.'

What followed was even more exciting. The leopard vanished into a cave, a markor appeared, dislodged a stone outside the cave, the leopard emerged 'almost like a spider coming out of its web', stalked and killed the markor and

# haunt of the piranha
## diving in the world's largest freshwater wetland

BRAZIL

● Pantanal

**BOLIVIA**

The piranha is the stuff of lore and legend. Step into the waters of the Amazon, dally too long and, before you know it, you've been ripped apart, your flesh torn from your bones in a bloody feeding frenzy that makes even great white sharks look a little dilatory.

Most biologists think such accounts are grossly exaggerated. But a prime-time television audience can be assumed to have a weak spot for fishy tales – and to be fair, you can never really be absolutely sure. Thus it was that *Planet Earth* began its quest for the piranha – and a team found itself flying over a wetland the size of England looking for a fish the size of a banana.

The Pantanal is a giant patchwork of lake, lagoon, river, forest, stream and swamp straddling the borders of Brazil, Paraguay and Bolivia. In and around its

waters live not just piranhas but also anacondas, caimans, killer bees – all animals with a certain fictional film pedigree. It doesn't sound the ideal place to go diving – and night-diving, at that.

Why piranhas? In a 50-minute programme, says producer Mark Brownlow, it's possible only to touch on the 'strongest characters', the most novel behaviours, and the piranha, because of the mystique that surrounds it, was a contender on both grounds. Moreover, piranhas have in the past been filmed mostly in aquaria. Filming them in the wild – notably in a type of habitat, freshwater wetland, largely ignored by underwater wildlife film-makers – was another opportunity to break the mould. And of course, the Pantanal, at some 165,000km² (64,000 square miles), is the world's largest freshwater wetland.

It's also a temperamental one, reliant on seasonal rains. The *Planet Earth* team, which included veteran underwater cameraman Peter Scoones, was primed to leave in March 2004, but the trip was called off at the last minute because of late rains and rapid drying. 'There's a narrow window just after the rains when the water levels are still high but the sediment has had time to settle,' explains Brownlow. 'That gives you a two- to three-metre [ten-foot] band of clear water. We had to be spot on in our timing.'

The go-ahead came in March 2005 from the team's guide in the Pantanal, the photographer Haroldo Palo Jr, and a six-person crew took up station in a local cattle ranch. The equipment, more than half a ton of it, followed by boat.

The Pantanal sequence in the freshwater programme features a cross-section of life in the waterworld – from fruit-feeding fish to the caimans that wait below stork and spoonbill nests to eat the chicks that fall out. But it was the underwater sequences that were the most ambitious – and their guide wasn't exactly dispelling the myths.

*Above*: Caimans lurked below birds' nests, waiting for the hapless chicks to fall. At less than 3m long, the reptiles were presumed too small to hunt people, though they managed to chew Peter Scoones' diving hood.

'When they begin to circle, keep moving, because if you're still for too long, they think you're dead and could take a nip out of you ... three days later we hadn't found a single piranha.'

'Haroldo gave us all this advice about filming piranhas. Don't go too far under thick mats of weed, because you won't be aware of where they are. When they begin to circle, keep moving, because if you're still for too long, they think you're dead and could take a nip out of you.' His caution seemed unnecessary, however. 'Three days later, we hadn't found a single piranha,' says Brownlow.

Water levels were the problem, they decided – too high, allowing the fish to disperse. So they began the first of many flights around the Pantanal, looking for likely piranha haunts – areas of clear water that appeared to be 'slightly lower, where more fish would concentrate'. Palo also made enquiries with fishermen. A new site was pinpointed. The team shifted quarters, flying to another ranch, the gear following by boat.

In all, they tried five different bases. 'We travelled the length and breadth of the Pantanal,' says Brownlow. At the second site, they found piranhas, but the fish, belying their reputation, seemed nervous, darting off under the weed. They tried

*Opposite*: Just after the rains, with clear water several metres deep, conditions were ideal for filming fish. But the ones the team really wanted to film – the legendary piranhas – proved very hard to find.

To get rare footage of piranhas under water in the wild, the team travelled the length and breadth of the Pantanal. They were on the lookout for slightly lower levels of clear water, where more fish might concentrate.

Filming from the surface as well as above and below it, the team aimed to capture a cross-section of life in a freshwater wetland – a habitat that few underwater wildlife film-makers had focused on before.

filming at night – no piranhas, again, but plenty of caimans.

'Far from being skittish like the piranhas, the caimans were inquisitive,' says Brownlow. 'They ended up chewing Peter's diving hood. Sharks will perhaps take a bite to see what you are and then leave you alone, but crocodilians actively hunt people in water. The caimans were impressive, but we gauged they were not big enough to size us up as food. The rule of thumb is that, if they're under three metres (10 feet), you can film them. Over three metres, they will hunt you.'

'We got awesome images of caimans swimming up to the camera, mouths open, but it was one of the eeriest dives I've ever done in fresh water. You can see what's in front of you only when you shine a light on it. And the mosquitoes cluster round anything exposed – your hands, your face – so you can't wait to get in the water.'

They finally caught up with a feeding frenzy after three weeks of trying. 'We saw some fish activity ahead, boiling and threshing, and so we dived in. There were hundreds of piranhas feeding on an injured fish. They never showed any aggression towards us. Peter doesn't use gloves to film in, and he was very casually brushing them away from his hand. By the end, we were free-snorkelling with them without wetsuits, just swimming trunks and bare flesh.'

The conclusion? That piranhas are not the voracious man-eaters of legend. 'We suspected that to be the case all along,' says Brownlow. But there's a caution attached. Many people they talked to in the Pantanal had been bitten by piranhas, usually on their fingers and as a result of chance or accidental contact. And much may depend on conditions. Suppose you lay limp in the water for several hours? Or suppose there was a drought, the pools were smaller and the piranhas had run out of food? In that case, Brownlow admits, things just might be different.

# hot spots and deep trouble

Diving deep inside the most treacherous caves or living on an active volcano in temperatures up to 50°C – there was nothing they wouldn't try. It was all about pushing the boundaries in the world's most extreme environments.

  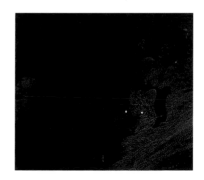

*Opposite*: Caving in Borneo – awaiting one of the most spectacular bat fly-outs in the world. *Left to right*: Building a microlight – getting off the ground in Mongolia. Masks ready – preparing to live on a desert volcano. Cat's eyes – finding a very rare leopard.

# the desert of death
## descending to the cruellest place on earth

DALLOL
Danakil Desert ●
ETHIOPIA

Wildlife film-making can sometimes be the stuff of nightmares. Imagine what it's like to sleep on a volcano, without a tent, on ground needle-sharp with solidified lava, feeling the warmth of the molten rock welling up from below. Or imagine lying down at night surrounded by the stench of sulphurous gases, wondering if you should go to sleep with your gas mask on and whether you will wake up again if you don't. Finally, imagine solifugids.

Solifugids are large, spider-like nocturnal creatures that make arachnophobia sound entirely logical. They emerge at dusk from their hiding places, 'twitter' by rubbing their jaws together and move fast and constantly. Among many names – hunting spiders, camel spiders, wind scorpions – they are also known as haircutters, because it's said they will munch their way out of human hair if they become entangled in it. That aside, they're harmless to humans – but when you're sleeping out in the open with no tent and lots of solifugids, that's probably not much comfort.

The ten days a *Planet Earth* team spent filming in northeast Africa's Danakil Desert in 2003 were among the most uncomfortable those taking part had

experienced. 'I don't think anybody slept a wink the whole time we were there,' says producer Vanessa Berlowitz. Partly it was the extremes they were facing – the Danakil is one of the hottest and lowest places on Earth; it's where the Earth's crust thins to near-vanishing point. But the solifugids didn't help. 'The lava was crawling with them,' says Berlowitz. 'You could hear them scuttling the whole night. It was horrible. You were constantly thinking they were going to crawl into your sleeping bag.'

It's hard to imagine a more inhospitable landscape than the Danakil. It's a baked and salty moonscape that has been described as the cruellest place on Earth. Its winds are scorching, its waters virtually non-existent and its resident tribes not kindly disposed towards outsiders. But the Horn of Africa is one of the most tectonically active regions in the world – 'you can actually see mountains being born there,' Berlowitz points out. For *Planet Earth*, it was thus primary viewing.

To film the business of Earth-building, a four-person BBC team – two cameramen, a researcher and a producer – travelled from the Simien Mountains of northern Ethiopia to the volcanic springs at Dallol and then on to the volcanic crater of Erta Ale, home to one of the world's four lakes of 'living' lava. It was a trip from the roof of Africa to one of its lowest points – the Danakil depression is 120m (390 feet) below sea level – and it was chosen because all three destinations lie along the huge tectonic fault line that has created Africa's Great Rift Valley. Here, two tectonic plates are being slowly pulled apart to reveal the geological crucible below. 'It's like a zipper that's being unzipped,' says Berlowitz. 'Where the zipper opens, the lava is oozing out.'

The poisonous landscape at Dallol Springs is strewn with colourful mineral deposits. It changes constantly: in temperatures of up to 50°C, new fumaroles emerge to spit boiling water and release volcanic gases.

*Above*: Nocturnal arachnids up to 55mm (2.2 inches) across, solifugids have powerful jaws to hunt small prey such as lizards. Camping in the open with them scuttling all around, it was easy to imagine them inside your sleeping bag.

*Top*: At Erta Ale, the crew camped inside the volcanic crater on old lava flows. The red-hot volcanic winds were so strong that it was impossible to put up tents – a few canopies provided the only shade.

Ten thousand years ago, the Danakil was part of the Red Sea, but mountain-building cut it off, and the water evaporated to salt. Now its daytime temperatures reach 50°C (120°F), enough in itself to make film-making problematic – even without living and working on a volcano or two.

'The Ethiopian army airlifted us to Dallol, slap bang in the middle of the Danakil,' says Berlowitz, 'but they didn't want to hang around, because they were worried about the temperature and the volcanic gases. So we had to jump out of the helicopter, and they threw down all our gear and took off. We were hit by this wall of unbelievable heat. We were operating in temperatures of more than 40°C [105°F], and we had to take all our own water in. We took ridiculous amounts – you worked out what you'd normally consume in a day and then tripled it.'

The temperatures made film-making an exercise in rehydration as much as camera angles. 'You could only walk about 50m [165 feet] before you were panting and had to lie down and drink some water. Every shot that in a normal environment would have taken a couple of hours took at least a day. You couldn't just knock a shot off – it had to be really carefully planned. By the time we got to Erta Ale, we were all suffering from heat exhaustion. Because it's so hot, your appetite vanishes – you're reluctant to eat because it makes you thirstier. So we concentrated on rehydration and dissolved glucose, which keeps your energy levels up. All of us lost a lot of weight.

'At Dallol Springs, we had to set up camp on a surface of crusty volcanic crystal. There were acid pools all around, it stank of sulphur and there were red-hot volcanic winds. You couldn't put a tent up, and so we had to roll out a mat and weigh it down with bits of lava. There was absolutely no privacy. At Erta Ale, we had to camp on the edge of the lava lake and wear gas masks or at least have them near us the whole time we were inside the volcano. We couldn't erect tents there either – we just lay down on blow-up mattresses on twisted old lava. I have vivid recollections of trying to sleep with the overwhelming smell of sulphur and wondering if I was going to wake up or whether this was the moment we were all going to be asphyxiated and then overcome with lava.'

Deserts, at night, are normally intensely cold. Not at Erta Ale. The Earth's crust there is said to be only 50m (165 feet) thick, and 'so not only were we camping inside a volcano, but it also stayed warm underneath the whole night. There was a weird sense of a boiling cauldron of activity only metres beneath you as you slept.'

'There was absolutely no privacy ... we just lay down on blow-up mattresses on twisted old lava, with the overwhelming smell of sulphur, wondering if we were all going to be asphyxiated ...'

Both crew and equipment spent most of the time during filming at Erta Ale clipped and roped to metal poles anchored around the sides of the lake. The lava in the lake, which was 50m (165 feet) below the lip at the time, had undercut the edge, and so filming down into the basin involved walking, securely roped, on a thin crust of old lava. Even so, it's hard to plan for every possibility. Italian geologists who routinely monitor the crater reckoned it had been constantly erupting and the lake overflowing for 100 years. The last recorded overflow was 60 years ago.

'The first moment I looked over and saw this bubbling magma was just amazing,' recalls Berlowitz. 'At night, when it glows orange and you can actually see the waves of lava, it was even more exciting. We all became obsessed with watching it.' A few weeks after the team's visit, the lava rose to the top. 'If we had been there when that happened, we would have been wiped out,' says Berlowitz. 'There would have been absolutely no chance of escape. We thought very carefully

Using a jib arm, the crew suspended a 35mm camera over the undercut crater lip to photograph the lava lake 50m below. A few weeks later, boiling lava flowed out over the spot where they had stood.

*Opposite*: Inside the crater, the crew kept gas masks on or within reach at all times. The air stank of sulphur, and suffocating fumes poured constantly from the volcano on winds that could change direction any moment.

Peering into the crater, the team watched the bubbling magma of Erta Ale, one of only four 'living' lava lakes in the world. The volcano boasts Earth's longest continuous eruption – so far lasting more than 100 years.

beforehand about safety procedures and the likelihood of eruptions. You do your best to calculate what's safe, but ultimately you are dealing with forces of nature.'

And then there were the hostile tribes. Compared with the chances of being caught in a lava flow, the dangers represented by Afar testicle-hunters might seem rather overdone. The Afars, nomads who inhabit the Danakil, have a reputation for fierceness and hostility towards outsiders: the men sport swords and rifles and are said to castrate their enemies and hang the testicles round their necks. How much of this is legend is hard to say. Nevertheless, the threat was considered sufficiently serious to merit stationing ten security guards armed with Kalashnikov rifles around the *Planet Earth* encampment.

It's probably no exaggeration to say that the sequences at Dallol and Erta Ale came spectacularly, if at times nightmarishly, close to the heartbeat of the Earth. Erta Ale will overflow again, and Dallol Springs will eventually emerge as a full volcano rather than a poisonous landscape of fumaroles venting hot gases and sulphurous springs. At Dallol, near where the crew spent their sleepless nights, there are the remains of what is thought to be a Second World War Italian potash-mining village. Warped by volcanic action, later bombed as a rebel hiding place, it now presents a strange spectacle of melted houses and old Fiats undermined and disembowelled by fumaroles – as though, says Berlowitz, 'the ground was reclaiming it and humanity was being wiped out.'

She adds: 'I have spent a lot of time on mountains, and there you get a similar feeling of being small and dwarfed and powerless. The lava in the lake at Erta Ale was a mirror of what was happening to the Earth's crust in the whole region. In Danakil you saw geology taking place in front of you.'

## sore point

The cinébulle is a relatively new concept in filming – a cameraman on a platform attached to a hot-air balloon, which can drift low over the ground giving an almost lyrical view. But when *Planet Earth* filmed cacti in the Sonora Desert, Arizona, wind, rain and flash floods almost put paid to the poetry. The climax came in a tug of war, with the crew yanking desperately on the balloon's ropes to stop it careering into a cactus forest. They failed – and the balloon was impaled on a giant cactus. The verdict? Great footage – eventually – but maybe cinébulles and cacti don't mix. Not when it's windy.

# shadows in the snow
## finding unexpected company

**Kedrovaya Pad Reserve**

**SIBERIA** ●

**VLADIVOSTOK**

When producer Mark Linfield set out to depict the seasons in a 'typical' deciduous woodland, he wanted to surprise viewers – to lull them into a sense of security, then produce something they didn't expect. 'We thought we'd put a twist in it,' he says.

To portray an English woodland, *Planet Earth* went to Siberia. Cameraman Barrie Britton and the team made two trips to the remote reserve of Kedrovaya Pad, north of Vladivostok, in 2004. 'We were constructing a sequence to look exactly like a woodland you'd find somewhere like Bristol,' explains Linfield. 'It looks very benign; you see many of the same birds, the same flowers, the same trees as you're used to – you expect to see a red fox or something similar. Then a big cat appears.'

The big cat in question is the Amur leopard, the most endangered subspecies of leopard but not one familiar to television viewers. There may be fewer than 40 left in the wild; by comparison with its neighbour, the Siberian tiger, it's rarely filmed. The two species overlap in range, but it's said that tigers kill leopards, which is why they're seldom found near each other.

To help with the shoot, Linfield enlisted the 'hard man' of Russian leopard filming, Anatoly Petrov. Petrov's style was distinctive – he would rig up a hide-cum-tent in the trees above a favoured leopard clearing and remain there for a week at a time. Explains Linfield: 'He takes everything up there with him – pots and pans, food, buckets for a toilet. He had a platform scarcely big enough to lie down on and stayed up there in temperatures of -25°C [-13°F] waiting for a leopard to show.'

Petrov's aim was twofold: to avoid leaving his scent trail on the ground and to make the most of leopard availability. He rigged up an early-warning system – infrared security beams along the main leopard-approach paths. When the cats broke the beam, they triggered a tiny blinking light on his platform.

Amur leopards may not have long left in the wild. They're hunted for skins and body parts, particularly by poachers who cross the border from China. Their lack of profile has also hindered conservation efforts. *Planet Earth* nevertheless filmed a mother and her cub feeding near Petrov's hide. And there was an undeniable twist to the filming – though not one that figured in the scenarios.

Unlike Petrov, Britton took time out from the hide. During a break, he was looking at tracks in the snow and found, near his own of a few hours earlier, a footprint 'the size of a dinner plate'. It belonged to a Siberian tiger.

Says Linfield: 'There was a deer farm not far away, and the tiger had shown up for an easy meal. It happens every now and again, and the government sends in a van of tiger-catchers. For a week, we had four guys in a truck with tranquillizer guns parked in our patch ... We never saw the tiger, and they didn't find it.

'Barrie was a bit shaken in an understated way. Walking through a snow-covered forest lays out all the natural history for you – whatever has gone on is written in the snow – so to suddenly find a Siberian tiger's footprint superimposed on your own is quite sobering, particularly in a forest that looks almost like a bit of English parkland.'

*Above*: Anatoly Petrov waited four days in the hide in freezing conditions for the first sight of this female.

*Top*: A fallen log may see regular use as a leopard bridge – so it was an ideal spot for a camera trap.

*Left*: The leopards were unperturbed by the hide and the cable linking it with the infrared beams below.

# waves in the sahara
## putting the zip into a sandstorm

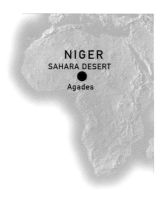

**NIGER**
SAHARA DESERT
Agades

Sensible people who live in the desert take cover when they see a sandstorm approaching. They hide under a rug, dive into tents, close most things that are open – their eyes, for example. Most wouldn't dream of photographing it.

A month or more after a stint in the Sahara trying to capture the unique visual qualities of a sandstorm for *Planet Earth*, cameraman Warwick Sloss was still scraping sand out of his ears. And his hair. And his equipment. His film camera had needed a pricey repair job. And he continues to marvel at the abrasive force of the sand. Some of the glass on the camera housing had been so badly pitted, you could barely see through it any more.

Filming sandstorms is not unlike waiting for a London bus. 'You can never catch up with them', says Sloss, 'and so you end up driving around the desert trying to find a position that will give you a good shot. Then you go out at the right time of day – mid-afternoon – in the hope that a good one comes along.'

You'd think that would be easy – sandstorms aren't exactly camera-shy. In fact, the team had two attempts, in 2004 and 2005, and both proved a challenge.

Explains Sloss: 'We filmed about eight or ten, of which only one was any good in terms of the way it looked. When they hit, you don't see a single wall of sand. It's more like mist rolling in – thin, misty, foggy cloud. It's very hard to get them to look nice on camera. You're looking for waves of sand coming in. You need the light hitting at the front or the side rather then behind, because otherwise you don't see any detail.'

Technical problems with film cameras hampered the first shoot, and so on the second, the crew tried digital still cameras running at one frame a second. The digitals could be sealed up in a plastic, waterproof-style housing. And when the film is played back at 25 frames a second, a speed with which the eye is comfortable, the sandstorms have a little more zip. 'If it's, say, a mile to the horizon and the sandstorm is coming in at 5mph, it's going to take a while for it to arrive,' says Sloss. 'It's a matter of compressing it into a more interesting timescale.'

Filming near Agades in Niger, Sloss developed a healthy respect for the desert. 'When sandstorms hit, it's pretty scary. Even if the sun is out it gets dark, almost like twilight, and everything turns orange. It's a bit like walking down the street under sodium streetlights. And everything fills up with sand – your eyes, your mouth, your clothes, the car, all the equipment.

'One day we found a sandstorm ... before it hit we jumped back in the car and stayed there, wondering how long it would take to go away. It didn't. It started raining. We were 10 miles [16km] from camp, and we drove back during what was now night. There was no visibility, the car was getting stuck what seemed like every 30 feet [10m], and we had very little idea of exactly where we were. There was water everywhere, and what had been desert just before was now lakes and lagoons. We finally made it back to camp and discovered that all our tents and camera boxes were under three feet [1m] of water. We spent two days drying and cleaning it all. I have never had an experience quite like that.'

**Cameraman Justin Maguire saw four sandstorms in the three weeks spent hunting for them in the Sahara in 2004. Despite problems with the film at daily temperatures of 43°C (109°F) in the shade, he got some dramatic footage and laid the basis for a second trip.**

# the great migration
## pushing for new views on the plains

**MONGOLIA**
● Choibalsan
ULAN BATOR

To anyone not involved in wildlife filming, some of the things film-makers do can seem bizarre. Who on earth would ship a microlight aircraft halfway round the world to film antelopes? Or bury a cameraman in the ground? So maybe it's not surprising when, sometimes, plans don't quite work out.

'The whole business probably sounds incredibly amateur,' says Jonny Keeling of his attempts to film the migration of the Mongolian gazelle, 'and I think, in some ways, it was. Most of the things film-makers are trying to do are new – no-one has done them before. So all you can do is work to the best of your experience.'

Keeling likens the steppe of eastern Mongolia to the 'biggest open field you've ever been in'. There are no trees, no hedges, no roads, very few lakes or rivers. It seems flat, empty and devoid of life. 'Open plains can look very dull, yet they have the biggest aggregations of animals anywhere in the world, certainly on land ... We wanted to get across what the plains are really like and maybe do this somewhere other than Africa.'

Most people have heard of the wildebeest or caribou migrations. Few, even in the academic world, know of the Mongolian gazelle's. Between one and two million of these small, brown antelopes commute yearly across the plains of eastern Mongolia between the Russian and Chinese borders – a distance of up to 1,000km (620 miles).

When Keeling went out to reconnoitre, he spent two weeks with field scientist Kirk Olson, driving round the open steppe. 'We had broken brakes, but it was probably about the only part of the world where it didn't matter. You go for miles

Cameraman Mark Smith spent a lot of time in body-sized holes dug in the steppe, with his head and his camera (covered in grass and camouflage netting) poking out of the top.

and miles and see nothing but grass and the occasional nomad. Finally, about four days into the trip, Kirk spotted a gazelle. By the time I got my binoculars out, it was a speck on the horizon, disappearing at full tilt. To say they were nervous would be an understatement.'

How do you film such a camera-shy antelope? Obvious, really – you bury the cameraman. Back in Mongolia the following year, Mark Smith spent a lot of time in body-sized holes dug in the steppe, with his head and camera (covered in grass and camouflage netting) poking out of the top. 'Mark was incredibly patient, and he did get some nice ground shots,' says Keeling, 'but the animals were very wary.'

It didn't help that the heat was intense – about 40°C (105°F) – that there was no shade and that the steppe was increasingly flammable. The concentrated fields of dung left by nomadic herders' animals are prone to smoulder into life

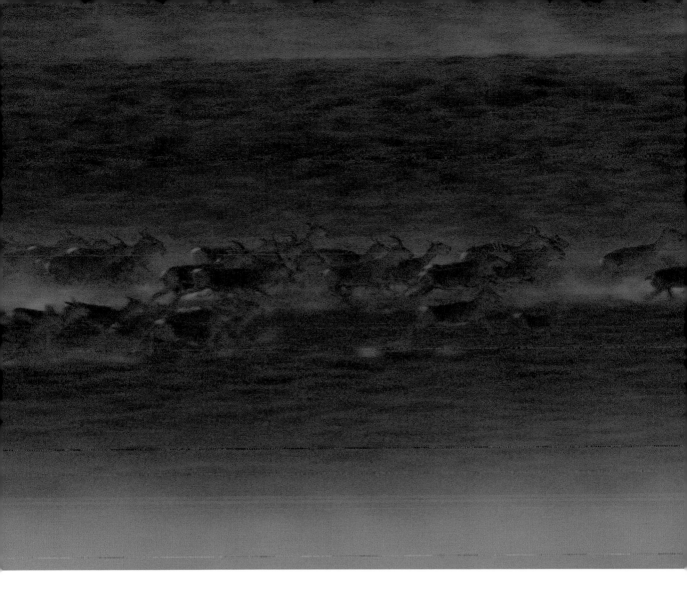

unpredictably. One did so while they were filming, destroyed a huge area of grassland and drove the gazelles away.

Keeling, Smith and researcher Emma Rolfe were camping on the steppe, getting up at 3am to go filming, coming back in the middle of the day when the best of the light was over. 'Sometimes we'd roll asleep under the car like panting dogs. Sometimes we'd get into the vehicle and switch it on to get some air conditioning. We could get ten minutes of comfort before we stepped out again.'

To convey a sense of the vastness of the migration in a land of unparalleled flatness, you clearly need aerial shots. Hence the next experiment – a microlight aircraft. Keeling reasoned that a helicopter would frighten the gazelles. But a microlight could fly high, cut the engine and drift down over the top of them.

The pilot chosen for the job was a Venezuelan named Hernando. First, however, his microlight had to be shipped from Venezuela to Mongolia. 'When it arrived, it was packed up in a huge balsa-wood case, which we jammed inside an old Russian van. It fitted between me and the driver to within an inch of the windshield. With various breakdowns and punctures and no paved roads, it took us three days to drive across country – all that time I was sitting with my head twisted to one side and the microlight case banging and rattling around.'

Filming conditions seemed ideal – beautiful light, fine weather and thousands of gazelles. But the animals were so nervous and the plains so open that it was a challenge to approach a herd without it fleeing.

Back on the plain, conditions were ideal – fine weather, good light and a herd of about 50,000 gazelles a few kilometres away. Hernando set about assembling the microlight the next day. Keeling liked his 'can do' attitude; he was less sure about the machine. 'A few alarm bells began to ring when Hernando showed me the on-off switch and said you had to switch it to 'off' to get the engine to come on.'

'We were still having breakfast when he jumped in, switched the on switch to off and started revving up the engine.' The next few moments are indelibly inscribed on Keeling's memory.

'The machine went whizzing across the steppe faster and faster until it started to wobble, snaked left to right, spun around and eventually hit a marmot hole. There was a huge puff of dust, the wheel flew off and the thing just flipped over onto its nose. The wings were folded and broken and completely torn.

'I dropped the camera and ran to him. There was a huge, unprotected wooden propeller at the back, and I had visions of him being chopped to pieces or burnt alive. The fuel tank had ruptured, and there was fuel coming out everywhere. The engine was hot, and we were surrounded by dry grass. The Mongolians who were helping us removed the fuel tank. Fortunately the pilot had only cut his knee. He walked off swearing and shouting, took off his helmet and smashed it on the ground, then sat down in tears, holding his head in his hands.'

The remains of the microlight are still in Mongolia – a local family made a fence and a gatehouse out of the aluminium poles. The cause of the accident was apparently a new wing, untested before its arrival in Mongolia. For the next few days, despite beautiful weather and gazelles to the horizon, aerial shots were out. Says Keeling: 'All that preparation and research and money, all the bureaucracy we went through, and in five minutes, the whole thing was shattered.'

In 2005, the team tried again. This time they mounted a camera on a bracket

The only respite from the intense heat was the shade of the vehicle (or a quick blast of its air conditioning). The crew regularly got up at 3am to catch the best of the light, and so by midday, they needed a break.

fixed to the outside of an old Russian military helicopter. 'We had to fly it from Ulan Bator to the far east of Mongolia. I was a bit apprehensive ... the machine stank of fuel, I was sitting on a huge additional fuel tank and, after about five minutes, I smelt something burning. I went into the cockpit to tell the pilot – and he was smoking a cigarette.

'The plan was to fly from Choibalsan airport to do the shots, going out at 5am, when the light was best. We'd spent ages getting the civil-aviation permissions. But at the last minute, the local official refused to come into work early.

'I was a bit apprehensive ... the machine stank of fuel, I was sitting on a huge additional fuel tank and I smelt something burning. I went into the cockpit to tell the pilot – and he was smoking a cigarette.'

'It wasn't as if they were particularly busy – there were only one or two flights a week from the airport, and air traffic control were playing table tennis a lot of the time. I actually played the guy and won – which probably wasn't very wise. Then I happened to mention it to someone quite important locally ... The next morning the official came in when we had asked. But we'd lost some lovely sunny days.'

The experiment succeeded, nevertheless. The helicopter hovered about 3km (1.8 miles) high, then slowly descended to 800m (2,600 feet) above the herds, by which time the animals had got used to the noise. For Keeling, the moral was about more than just technique: 'For these things to work, you need all the planets to be aligned perfectly, and the chances of that are slim ... but sometimes it does happen.'

It took six months' planning and logistics to ship the microlight from Venezuela to Mongolia. Quieter than a helicopter, it was meant to film aerials of the gazelles. But it crashed on take-off, leaving the pilot in despair.

# tales of the underworld
## from toxic caves to underground waterfalls

**MEXICO**
Villa Luz
YUCATÁN

A cave entrance leading to underground lakes and rivers is known in parts of Central America as a cenote – a word inherited from the Maya signifying an opening into the underworld. To the team making the programme on caves, the term sometimes seemed uncomfortably appropriate.

Caves can undoubtedly be beautiful places – with unique ecosystems and wildlife and their own haunting and magical settings. But they can also present something close to a vision of hell on earth – a scene that could have come straight from a painting by Hieronymus Bosch. And since several of the caves filmed for the series seemed to fall into the Bosch-like category, the logistical challenges facing the production team were extreme.

Probably the biggest challenge was posed by the 'poison cave' of Villa Luz in Mexico. The importance of this cave ecosystem was first realized by explorer-researcher Jim Pisarowicz in 1987. He found the strange globular mats dangling from the ceiling and walls fascinating if not necessarily charismatic – he christened them snottites – but he emerged to find his clothes beginning to dissolve and large red welts forming on his skin.

The Villa Luz cave is a miasmic vortex of toxic gases, acids and alkalis – even by the anoxic standards of the underworld, it is regarded as 'exceptionally poisonous', says assistant producer Kathryn Jeffs. Hydrogen sulphide wells up in great watery belches from underground springs and feeds thick, slimy mats of acidic bacteria on the walls and roof. The bacteria supply a constant drip of sulphuric acid; the mud on the floor of the cave, meanwhile, can be highly alkaline and caustic. Then there are the fraying walls, the collapsing ceilings, the swarms of buzzing gnats …

'The acid-based ecosystem is eating away the cave from the inside,' explains Jeffs. 'Often huge chunks of the roof would fall down, hitting the acid lakes below with loud, echoing splashes. We had to become experts on the effects of all the different gases, which could build up rapidly in lethal pockets. It's not a deep cave – it would probably take only 20 minutes to get back to the entrance – but the gases could be so deadly that you might have only minutes to react.'

The film crew wore protective clothing and gloves and full face-masks with goggles to stop gases being absorbed through the eyes. They also carried constantly beeping gas monitors to warn of drops in oxygen levels or sudden spikes of toxic fumes. Even so, during the ten days they spent in the cave, several of the crew felt dizzy or sickly, some developed red marks on their legs. And though none of these effects was long-lasting, cameraman Rob Franklin reckoned the acid had eaten away some of his lighting and filming gear and was convinced it had dissolved his spectacles.

A deep cave complex in the hill country of northern Thailand presented a different challenge. The target of this sequence was the 'cave angel' – a small but acrobatic fish, which lives around a waterfall several hundred metres below

*Opposite top*: Even near the entrance of the poisonous cave of Villa Luz, assistant producer Kathryn Jeffs had to wear a mask and gas detector. To go deeper, the team needed full face masks and protective clothing.

*Opposite bottom*: They also entered Mexico's remote Cave of Marbles, which contains millions of 'cave pearls' made of limestone. One of the chambers can be reached only by squeezing through this tiny gap.

*Opposite*: This warm, sulphur-rich stream led explorers Jim Pisarowicz and Warren Netherton to Villa Luz cave in 1987. The water's milky appearance is caused by the large amount of sulphur suspended in it.

*Left*: Some locations were so remote and flooded that they were accessible only by wading. It took about an hour twice a day to cross this marshland with all the kit – and a few large leeches usually attached on the way.

CHIANG MAI
● Lum Nam Pai
Wildlife Sanctuary
**THAILAND**

ground in a subterranean river system under a rocky massif near Chiang Mai. It's one of only two cave systems where the fish is found. It's also an environment with no light, varying levels of oxygen and an abundance of fast-flowing water and jagged, slippery, unstable rocks. The waterfall itself is half an hour's journey by kayak along the underground river from the nearest cave entrance.

'The cave was remote and much deeper and longer than the one in Mexico,' explains Jeffs. 'We had to kayak down a river, then carry the kayaks and kit through the forest to the cave entrance, where we made camp. We then had to lower the kayaks and all the equipment down a steep slope in the cave to the underground river. We had to half paddle, half drag the kayaks to get to the waterfalls. The oxygen levels sometimes dropped and there were a lot of obstructions, and so we were breathing hard. It was exhausting.

'To film in what is normally a pitch-black environment, we were working with high-power, expensive lights – not all of them were waterproof. Because of the noise of the water, we couldn't hear each other. So we were clambering about very wet, slippery rocks, doing lots of hand signals and trying not to let the lights fall into the water.'

They stopped the lights falling in but not the humans. The waterfall was impressive, about five metres (16 feet) high, but John Spies (local caving expert and consultant) knew there was another, bigger one upstream. It was even more spectacular and might also have resident cave angels.

'We had all had waterfall caving training and experience, and there were highly trained cavers with us, and so we started to climb the waterfall. Since there was no rope, John set off first. As he reached the top, I started climbing. The water was very cold and the rocks were sharp and slippery. When I got to the top, I was clinging onto a rock with one hand and supporting myself on another rock with the other. I was just about to heave myself over, and could see John reaching out to me, when the rock in my right hand came away.

'Luckily I fell backwards, pushed by the force of the water, rather than straight down onto the rocks below. I plunged into the pool and remember thinking that at least I hadn't hit any rocks. Then I felt the floor of the pool and pushed myself

'I remember thinking that at least I hadn't hit any rocks ... But I had a pack on my back, which filled with water and dragged me down again.'

Three days before the dive was to begin, an empty taxi was found by the cenote – and the body of its driver, weighted with engine parts, in the waters ... the filming location had become a murder scene.

Diving in a cenote and through an underground river system, the team swam past algae and tree roots. The river sustains the forest above, with roots pushing down through the limestone to reach the fresh water.

back up to the surface. But I had a pack on my back, which filled with water and dragged me down again. I pushed back up a second time, and as I reached the surface, I was incredibly relieved to feel hands grab hold of me and drag me out. It was Justin [Maguire, cameraman], who had been ready to leap into the water straight away.

'The first thing I said to him was: "What are you doing here?" He was our most precious resource – the cameraman. Not thinking straight, I was annoyed that he had had to jump in.' But the fact that everyone was ready to deal with the situation and that a more serious incident was avoided was a reflection of the intensive preparation and training the team had put in before the trip.

Back in Mexico, the flooded caves of the Yucatán gave an altogether grimmer taste of the underworld. The toughest challenge of the shoot was to film remipedes, ancient underwater predators that look like swimming centipedes and feed on cave shrimps. Remipedes were abundant before fish evolved but now

survive only in deeper cave waters inaccessible to most fish. The dive would involve drops down several long, flooded passages, each one narrower and steeper than the last.

Three days before the dive was to begin, an empty taxi was found by the cenote – and the body of its driver, weighted with engine parts, in the waters. Worse, a woman's purse was found in the taxi – leading police to suspect that she, too, had been murdered and dumped in the cenote. The filming location had become a murder scene – with one body, apparently, yet to be recovered.

As it turned out, the woman was found unharmed. It was rumoured that she was the girlfriend of a local gangster who was languishing in prison: she had taken up with the taxi-driver, and her boyfriend, on hearing this, had sent two of his men to deal with his rival. The dive took place as planned.

The underworld left the crew with some positive memories, however. 'The Thai waterfall cave especially was thick with midges,' recalls Jeffs. 'They were attracted by your head-torch; they swarmed around your face, got in your mouth and up your nose, you breathed them in. It could be unbearable. Lunch became an exercise in trying not to eat flies. But funnily enough, it eventually became my favourite time down there – we learnt to turn off all the lights as soon as the pack lunches were handed out and to sit eating quietly in the pitch black. The calm and peace just sitting in utter darkness, hundreds of metres underground, listening to the pounding waterfall was amazing.'

**The formations in the crystal-clear waters of the Yucatán flooded caves have changed little in thousands of years: created by dripping water, the stalactites stopped growing when the caves were flooded.**

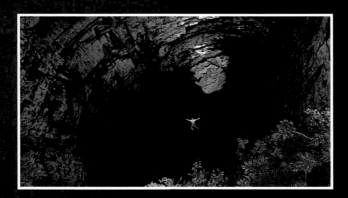

Mexico's Cave of the Swallows – named after the thousands of swifts that roost there – is the deepest cave hole in the world, plunging more than 400m (1,300 feet). Even so, the base-jumpers filmed by *Planet Earth* to give a sense of scale had just seven seconds in which to open their parachutes.

# the eve of the swarm
## performing tricks in mauritania

**NORTH AFRICA**
● **Nouakchott**
**MAURITANIA**

There's a small corner of West Africa where *Planet Earth* – and in particular assistant producer Tom Hugh-Jones – will be remembered with particular awe. Quite what label they will attach to Hugh-Jones is hard to say – though The Man Who Catches Vomit is a strong possibility. Legendary status, nevertheless, seems assured.

Given the all-consuming reputation of the locust and the damage it inflicts on crops, you'd think a locust swarm would be easy to film. But when Hugh-Jones and cameraman Brian McDairmant went to Mauritania in 2004, it didn't work out that way.

To begin with, a television series and a national locust control agency may have different ideas about what constitutes telegenicity. The first tip that the locusts were on the move came from locust-control experts in February 2004, but it turned out to be nothing on the scale the film-makers were looking for. 'There was the odd tree with locusts in it and even quite a few in the grass, but that's very unimpressive when you film it,' says Hugh-Jones. 'You really need a huge density of animals.'

Film crews also have different agendas from locust agencies (and farmers). Most Mauritanians were happy that the locusts hadn't arrived, for example. And since the *Planet Earth* team were reliant for tips of swarming behaviour on locust controllers, they might turn up only to find the locusts sprayed and dead. That's without reckoning on the locusts themselves, which moved so fast that the camera team would constantly be playing catch-up – arriving to find the insects had already departed.

'We asked her if it was polite to burp in Muslim countries, and she said, yes, it was a way of showing you're thankful for a good meal. So I thought I'd show them my trick.'

The first trip was thus utterly miserable, pursuing non-existent locust swarms across Mauritania. 'We were in the car for a good eight hours every day, doing maybe 300 miles [480km] a day. We were basically driving non-stop around the country all day every day. Everybody got very despondent.'

At this point, Hugh-Jones did what any Englishman up against it in foreign climes would do – he decided to boost his troop's morale. 'We'd all sit on the rug in the tent in the evenings. We'd play a game or try to entertain each other. One night after dinner, we were sitting round in a circle, and our cook did a huge burp. We asked her if it was polite to burp in Muslim countries, and she said, yes, it was a way of showing you're thankful for a good meal. So I thought I'd show them my trick.

'The people there were helpers – the locust control agency, guides, cooks – all devout Muslims. My trick is that I can 'breathe' air into my stomach and then let it out again in a very long and loud burp. I explained this to them in French, and I don't have the best French, and so they were all wondering what I was going to do. I started to do a huge burp, but it brought up the food I'd just eaten in a spout of vomit. In my surprise, I reached my hand out and caught the sick in it.

'Everyone just looked at me in shock. Basically they thought that was my trick. I didn't have the French to explain it wasn't. Then they screamed and got up and ran off. They didn't come back for about five hours. They thought I was really disgusting.'

It was a story with a happier ending, however. Hugh-Jones and cameraman Mike Potts returned to Mauritania seven months later and found the spectacle they wanted – a swarm of locusts that 'started about 2pm and was still flying overhead at 7.30pm when we left'. The sequence forms a remarkable ending to the programme on deserts. Hugh-Jones still cringes at the memory of his first visit to the country but thinks his audience eventually forgave him. 'For the rest of the trip, they thought I was really weird,' he says. 'But I think they found it funny in the end.'

*Left top*: Driving through a swarm was like being in a hailstorm, except the vehicle was plastered with locusts.

*Left bottom*: Each night, the team camped in the nearest dunes, sometimes sharing the site with the hospitable nomadic pastoralists.

*Below*: Finally, the crew caught up with a massive swarm, estimated at 65km (40 miles) wide, which flew overhead for more than six hours.

# home base
## making it all happen

*Above*: Based in Bristol, the production coordinators – (left to right) Susan, Lisa, Joanna, Sarah, Samantha and Amanda – faced some of the most challenging preparations ever needed for a natural history series.

Almost every mode of travel was used, including (*clockwise from top left*): a microlight in Mongolia, quadbike in Gabon, van (to carry the microlight), helicopter, vehicles in Egypt and a canoe in Gabon.

Picture a job that involves planning and provisioning an expedition, shipping bulky and bizarre items from one corner of the world to another, sending sensitive equipment and people into extreme environments, and then making sure everything meets the toughest health and safety rules, and you will get the flavour of a production coordinator's role on *Planet Earth*. It's challenging work at the best of times. Occasionally it defeats you.

Sarah Wade thought she'd hit upon the answer to personal hygiene problems in the plunging winter temperatures of the Gobi Desert in Mongolia, for example. It was too cold to take your clothes off to wash, even supposing there was any water, and so the crew travelling to the Gobi for two months to film wild Bactrian camels were supplied with wet wipes – hundreds of them. 'They all froze solid and were a total waste of time. The crew came back stinking to high heaven.'

Then there were the urine bottles, for those tricky nocturnal moments. Wade had the foresight to provide them, for an all-male team, but they were made of plastic. If you weren't careful when defrosting them over the fire, the plastic melted. 'I only got them one bottle each, which turned out to be a bad plan,' she says. 'I'm not sure how they managed after that.'

In wildlife film-making, particularly for a landmark mega-series, the weeks spent on location are the tip of the iceberg. Behind it lies months, sometimes years, of research, planning and preparation – much of it trying to predict the unpredictable.

How do you ship a microlight from Caracas to Mongolia, for example? (Via Berlin, naturally.) What about transporting a hot air balloon from Lille to Gabon? How do you get a jimmy jib (a mini-crane that carries a camera) onto a Venezuelan mountaintop? Or a ton of lighting equipment into a cave in a forest in Borneo? Where do you find a paper boiler-suit for a producer up to his neck in bat excreta? How do you find somebody in Vladivostok who speaks English? (You stay at the office till 11pm and try the American consulate.) And how do you make sure everybody eats properly?

That last one is easy, by the way. 'You consult the BBC's safety store in London, which supplies a calorifically calibrated dried-meals package. And then you march off to the supermarket and buy the extras – nuts, snacks, cereal bars,' explains Joanna Verity. 'The food was a bit grim in the Gobi,' recalls Wade. 'You just can't get fruit and vegetables out there. I got kilos and kilos of chocolate, masses of dried fruit, mega-vitamin supplements. They took suitcases full of food supplies.'

But the BBC is a house of many rooms. There are shipping experts, health experts and even experts who can tell you which gas masks to use in a cave or a volcano. In the safety store, there are life jackets and hard hats and crash helmets and trauma kits. In Bristol, meanwhile, there's the *Planet Earth* 'kit cage', an Aladdin's cave of climbing ropes, caving gear, diving equipment.

'There's nothing they won't try. They're wanting to do everything that's new, that hasn't been done before – and a lot of the time, it hasn't been done because it's nigh-on impossible.'

Home base thus has an expeditionary flavour. 'There are maps on the walls, postcards from all over the world,' says production team assistant Samantha Davis. 'And there's kit everywhere. You'll see them testing walkie-talkies ... and putting up tents in the office to make sure there are no holes in them or to find out if three people can fit in, though it's only meant for two.'

The production side of *Planet Earth* was 'extremely demanding' – the verdict of series producer Alastair Fothergill. That has a little to do with the notorious unpredictability of wildlife filming and a lot to do with the ambition and scale of the undertaking. It took a year to persuade just the BBC to let the series film in the Karakoram mountains of Pakistan, such were the perceived political dangers.

Nearly 30 staff – producers, researchers, coordinators – worked on *Planet Earth*, and they've been at it since 2003. Another 30-40 cameramen have been involved – and beyond them hundreds, quite possibly thousands, of scientists, guides, field assistants, helpers. For those at the centre, it has been something of a roller-coaster – up when sequences worked, down when they haven't.

Wade calls it 'quite the most challenging production I've ever worked on. It's incredibly exciting and absolutely exhausting – a real test of strength.' Why? 'Because there's nothing they won't try. They're wanting to do everything that's new, that hasn't been done before – and a lot of the time, it hasn't been done because it's nigh-on impossible. But if you've got the time and the budget, you just stick at it and make it happen.' Or as Verity says: 'It's not impossible, because we did it.'

Imagine the hundreds of permits, carnets and visas required to send the film crews and their kit to all corners of the globe – then add the preparations needed for filming in the most extreme places on Earth.

# credits

**PUBLISHED BY BBC BOOKS**
BBC Worldwide Ltd
Woodlands
80 Wood Lane
London W12 0TT

**First published 2006**
**Reprinted 2006 (six times)**
© BBC Worldwide 2006

**COMMISSIONING EDITOR**
Shirley Patton
**EDITOR**
Jane Wisbey
**DESIGNER & ART DIRECTOR**
Traci Rochester
**PICTURE RESEARCHER**
Laura Barwick
**MANAGING EDITOR**
Rosamund Kidman Cox
**PRODUCTION CONTROLLER**
David Brimble

ISBN-10: 0 563 49358 5
ISBN-13: 978 0 563 49358 7

**ORIGINATION & PRINTING BY**
Butler & Tanner Ltd, Frome, England

## coming soon

To accompany the second part of the series, this lavish book will be published by BBC Books in October 2006.
ISBN 0 563 52212 7
£25.00 RRP

## picture credits

**FRONT cover** – Fred Olivier; **BACK cover top** – Ben Osborne; **BACK cover bottom** – Martyn Colbeck; **1** – Minh-Thu Pham; **2/3** – Sue Flood; **4/5** – Fred Olivier; **7** – Gavin Newman; **8** – Fred Olivier; **9 top** – Jonny Keeling; **9 bottom** – Huw Cordey; **10/11** – Tom Clarke; **12** – Jeff Wilson; **13 left** – Fred Olivier; **13 middle** – BBC NHU; **13 right** – Huw Cordey; **14/15** – Huw Cordey; **15** – Huw Cordey; **17 top** – Tom Clarke; **17 bottom left** – Huw Cordey; **17 bottom right** – Huw Cordey; **19** – Martyn Colbeck; **20** – Jeff Wilson; **21** – Vanessa Berlowitz; **22** – Jeff Wilson; **23** – Tom Hugh-Jones; **24** – BBC NHU; **25** – Jonny Keeling; **26** – Jonny Keeling; **27** – BBC NHU; **28/29** – BBC NHU; **30** – BBC NHU; **31 top** – Jonny Keeling; **31 bottom** – BBC NHU; **33 top** – BBC NHU; **33 bottom** – Justin Anderson; **35** – DCI/Martin Klimek; **37 top** – Mark Brownlow; **37 bottom** – Didier Noirot; **38** – Mark Brownlow; **39** – Mark Brownlow; **40/41** – Fred Olivier; **42/43** – Fred Olivier; **44** – Fred Olivier; **45** – Wade Fairley; **46 top** – Wade Fairley; **46 bottom** – Fred Olivier; **49 top** – Mark Brownlow; **49 bottom** – Richard Burton; **50** – Mark Brownlow; **51** – Richard Burton; **53 top** – Jason Roberts; **53 bottom left** – Jason Roberts; **53 bottom right** – Doug Allan; **54** – BBC NHU; **55** – Doug Allan; **56** – Magnus Elander; **57 left** – M & P Fogden; **57 middle** – Tom Clarke; **57 right** – Jeff Wilson; **58 left** – M & P Fogden; **58 right** – Tom Hugh-Jones; **59 top** – Morley Read/naturepl.com; **59 bottom** – Enigma/Alamy; **61 top** – Freya Pratt; **61 bottom** – Jonny Keeling; **62** – Jeff Turner; **63** – Freya Pratt; **64** – Jeff Wilson; **65** – Jeff Wilson; **66** – Charlie Hamilton-James; **67** – Eddie Gerald/Alamy; **68 top** – BBC NHU; **68 bottom** – BBC NHU; **69** – Mike Dilger; **71 main** – Mark Linfield; **71 inset** – Mark Linfield; **73 top** – Ben Osborne; **73 bottom left** – Ben Osborne; **73 bottom right** – Ben Osborne; **74/75** – Ben Osborne; **77** – Tony Wu/seapics.com; **78/79** – Sue Flood; **81** – Tom Clarke; **82** – Tom Clarke; **83** – Jeff Wilson; **85 top** – BBC NHU; **85 bottom** – Jeff Wilson; **86/87** – Jeff Wilson; **88 top left** – BBC NHU; **88 top right** – BBC NHU; **88 middle** – BBC NHU; **88 bottom** – BBC NHU; **90** – Peter Scoones; **91** – Peter Scoones; **92** – Peter Scoones; **93** – Doug Anderson; **95 top** – Jonny Keeling; **95 bottom** – BBC NHU; **96/97** – Magnus Elander; **98** – Ted Giffords; **99 left** – Jonny Keeling; **99 middle** – Minh-Thu Pham; **99 right** – Anatoly Petrov; **100/101** – Martyn Colbeck; **102 top** – Minh-Thu Pham; **102 bottom** – M & P Fogden; **103** – Martyn Colbeck; **104/105** – Martyn Colbeck; **105 top left** – Minh-Thu Pham; **105 top right** – Tom Hugh-Jones; **107 top** – Anatoly Petrov; **107 bottom left** – Anatoly Petrov; **107 bottom right** – Anatoly Petrov; **109** – Chadden Hunter; **111** – BBC NHU; **112** – Jonny Keeling; **113** – Jonny Keeling; **115 top** – Kathryn Jeffs; **115 bottom** – Kathryn Jeffs; **116** – Kathryn Jeffs; **117** – Henry M. Mix; **118** – Andreas W. Matthes; **119** – Gavin Newman; **120/121** – Stephen Alvarez/National Geographic Image Collection; **120 top** – BBC NHU; **120 middle** – BBC NHU; **120 bottom** – BBC NHU; **123 top left** – Tom Hugh-Jones; **123 bottom left** – Tom Hugh-Jones; **123 right** – Tom Hugh-Jones; **124** – Jeff Wilson; **125 top left** – Jonny Keeling; **125 top right** – Ben Osborne; **125 middle left** – Ben Osborne; **125 middle right** – Jonny Keeling; **125 bottom left** – Danny Cleyet-Marrel; **125 bottom right** – Jonny Keeling; **126/127** – Ted Giffords; **128** – Fred Olivier